DEVELOPING SMARTER

DEVELOPING SMARTER

A PATH FORWARD FOR COASTAL REAL ESTATE

AN IN-DEPTH STUDY OF THE INCREASING
RISKS ASSOCIATED WITH NATURAL DISASTERS
IN COASTAL REAL ESTATE COMMUNITIES

WINSTON SMITH

NEW DEGREE PRESS

COPYRIGHT © 2018 WINSTON SMITH
All rights reserved.

DEVELOPING SMARTER
A Path Forward for Coastal Real Estate

ISBN   978-1-64137-094-3 *Paperback*
        978-1-64137-095-0 *Ebook*

To Keiko, my childhood dog, who recently passed during the process of writing this book. Keiko was a Portuguese Waterdog and certainly lived up to her breed. Growing up, we shared a love and passion for the ocean and spent many summer days playing in the waves. It only seems right to dedicate a book about our coastlines to the lasting memories I have with her at the beach.

CONTENTS

PART I: HOW BAD IS IT?
INTRODUCTION 13

CHAPTER 1: THE EXTENT OF THE PROBLEM 29

PART II: HOW DID IT GET THIS BAD?
CHAPTER 2: THE FLOOD INSURANCE PROGRAM 39

CHAPTER 3: UNDERSTANDING THE EXPERTS 51

PART III: ENTERING A PROACTIVE MINDSET
CHAPTER 4: GOOD, FAST, CHEAP 67

CHAPTER 5: INSTITUTIONAL INTEREST 77

PART IV: MARKET CASE STUDIES
CHAPTER 6: HOUSTON WE HAVE A PROBLEM 91

CHAPTER 7: NEW YORK IS WORTH SAVING 123

CHAPTER 8: BOSTON PUSHES FORWARD 151

PART V: THE KEYS TO LASTING CHANGE
CHAPTER 9: THE BENEFITS OF PRIVATE INSURANCE 177

CHAPTER 10: BID, RELI, AND A CALL TO ARMS 193

ACKNOWLEDGEMENTS 209

REFERENCES 211

*We cannot stop natural disasters but we
can arm ourselves with knowledge: so
many lives wouldn't have to be lost if there
was enough disaster preparedness.*

—PETRA NEMCOVA

PART I

HOW BAD IS IT?

INTRODUCTION

"I expected the worst, but it was beyond any expectations I had. The destruction there was the most severe in the area." Amanda Lee, FEMA Volunteer

"It was all a nightmare waiting to happen." Chris Garwood, Vice President at DCS Design

"There are a lot of high-tech buildings in the area but most were still built without the consideration of climate change in mind." Paul Kirshen, Civil Engineer and Professor at the University of Massachusetts Boston School for the Environment

"If the angle of your book is to identify inefficiencies, there are plenty to pick from. One of the questions you have to ask is whether or not you can even squeeze out enough efficiencies to save the program." John Doe, a high-ranking official at a subcontractor for the National Flood Insurance Program

"People are saying, 'Let it go back to the beavers,' but New York City is worth saving." Bill Golden, President of the National Institute for Coastal and Harbor Infrastructure

"You can find online that Einstein defined insanity as doing the same thing over and over again and expecting a different result." Ed Thomas, National Hazard Mitigation Association Volunteer

"Even if you don't believe it," says Pina Albo, past president of the re-insurance division of Munich Reinsurance America Inc., "[natural disasters] happen, so how do you prepare yourself for future ones better than you did in the past?"

* * *

In late August of 2017, I was sitting at the airport in Providence, Rhode Island ready to board my flight. This airport, miles away from the Caribbean, was still buzzing about what we were all seeing on the news.

A category 5 hurricane, named Harvey, was on track to hit the Texas Gulf Coast, including Houston, Texas.

A number of my friends at the University of Miami were packing up and leaving school after being on campus for only a week. On television and social media, everyone was talking about when, where, and how hard it would hit when it made landfall.

But part of me wondered, why are we so surprised?

"We expect everywhere that flooding rains will increase," said Ken Kunkel, an official at the National Oceanic and Atmospheric Administration. This was back in 2015 in response to the deadly flooding in South Carolina. "You can build it now for today's climate, but you may not be quite in tune to what the climate will be in 50 years."

As someone who has grown up in the era of increasing climate change awareness and discussions, along with witnessing growing devastation, I wondered, "Why are they calling this a 500- or 100-year flood or hurricane when now these types of natural disasters seem to happen every other year?"

"People have for years and years wanted to live along the water," says Laura Lightbody, who directs the flooding program at the Pew Charitable Trusts. "There are some perverse incentives in place from a policy perspective to continue to build and live along those areas, which is making them more at risk."

And sadly, the damage was horrific. After getting hit by two major hurricanes in a matter of weeks—Hurricanes Harvey and Irma—Moody's Analytics estimated that damages to Texas and Florida cost between $150 billion and $200 billion, comparable to the costs from Hurricane Katrina in New Orleans in 2005.

$200 billion.

Decades ago, we would not be able to predict the expedited path and duration of a storm with the accuracy we have today. Look at how far we have come in predicting storms. With this technology, we have identified the most vulnerable coastal communities to incoming storm surges.

Experts like Kunkel and Lightbody aren't alone in predicting these catastrophic events hitting our major coastal cities. It is not that we couldn't have predicted this—in fact, experts have said it was a matter of *when*, not *if*, a lowland area like Houston would be hit with a tropical storm and that inevitable mass flooding would ensue.

If we can predict the storm's path and even forecast a conservative plan for the upcoming decade or decades, why have we not used this new information and exercised the same innovation in the development and infrastructure within these communities?

Historically, the real estate industry has gained a reputation for being a slow adapter of technology. This characterization of the industry is what may contribute to further damages if we do not change the way we develop coastal real estate communities.

If it is no longer a surprise, why do we keep letting these events leave us flatfooted?

As the *Time* cover story was aptly titled following Harvey and Irma:

"Why We Won't Be Ready for the Next Hurricane Harvey Either"

Ouch.

* * *

While there is a lot of scientific evidence surrounding climate change and the potential for rising sea levels, the purpose of this book is not to have a political debate about climate change but to address the real concerns facing the real estate industry. While much of the content will reference climate change and the impact of rising sea levels on coastal real estate communities, it isn't even necessary to believe in climate change to realize we have a growing problem. Even if the climate maintains its current state, development along coastal communities continues to increase. The growing value of coastal properties is one of the largest factors impacting increased hurricane damage risk. Around the world, coastal populations continue to grow faster than non-coastal areas because people simply love living by the sea.

We need to start building smarter along our coastlines before we lose them and the families that inhabit them.

We have to stop feigning surprise.

Wallace J. Nichols is infatuated with human's relations to the sea. And he's someone who is *not surprised*.

In his book "Blue Mind," he dissects our obsession with the ocean by focusing on every aspect of the sea that attracts us, from swimming, the ocean breeze and hot touch of sand on a sunny day, to the endless crustaceans and fish we consume, to the use of water as transportation, and by asking poignant questions such as "Why is 'ocean view' the most valuable phrase in the English language, bestowing a 50% premium on everything from lunch to a night's sleep in a hotel room to a beach front cottage?"

Globally, the ocean grants a trillion-dollar premium on hotel rooms, condos, houses and all other forms of coastal real estate. People are willing to pay the 'blue premium' in order to be close to the sea. It is this innate human desire to be close to the sea that has driven the rapid development of coastal shoreline communities and led us to the problem at hand. Fourteen of the world's seventeen largest cities are now located along the coast.

In the U.S. alone, coastal counties, which account for 14 of the country's 20 largest metropolitan areas, grew by 84% between 1960 and 2008 compared to 64% of non-coastal areas. In 2010, 40% of the country's population lived in coastal shoreline counties, and that number is expected to increase by 8% by 2020.

News Headlines like *"Florida's Overdevelopment Has Created a Ticking Time Bomb"* from *Scientific American* demonstrate the extent of the problem and the increasing risk that coastal communities face.

It's certainly a problem, but perhaps the right question to ask is: "Why isn't anyone doing anything about it?"

* * *

At this point you should be asking yourself why me? Why real estate? And why focus on this problem?

Well, I'm a nerd about real estate.

It's not a conventional career, and many of the most gifted real estate professionals I've interviewed for this book will admit they often 'backed' their way into the real estate career track.

But not me, I've always known I wanted to work in real estate.

On my sixth birthday, my grandmother gifted me an architectural design kit. It was a basic set for kids that included open floor plans, pencils, and traceable objects to fit within the floor plan. Back then, my kindergarten went from 8am to noon, and the afternoons I spent playing with that architectural design kit are my earliest recollections of my interest in real estate. I simply loved playing with the floor plans. It was always a puzzle to me. I would start with a square and figure out how to fit a kitchen, a bedroom or two, a bathroom, family room, etc. into the space. I would spend extended time periods internally debating the appliance locations within the kitchen or the direction that doors would swing to ensure I was using the space efficiently and the different areas all flowed well into one another. Sometimes, I would even challenge myself by starting with abnormally shaped spaces and designing them to flow effectively.

I'm sure my grandmother's gift idea came from her son, my Uncle Curt, who studied architecture all throughout college and into graduate school. He spent his early career as an architect before he became involved in development. His leadership, guidance and support allowed my juvenile interest to grow into my passion for the industry today.

After coming down with a severe stomach bug that put me in the hospital during college, I quickly reminisced about my elementary school days when I could stay home sick from school with my mother. Like most kids at that age, I enjoyed staying home from school, but likely not for the reasons you might suspect.

I would stay home to watch Home & Gardens Television or "HGTV".

See, I told you I'm a nerd.

Every opportunity I got to stay home, I would spend the entire day on the coach watching an endless lineup of HGTV shows. I was fascinated by all of them, from the house hunters to the house flippers. I would even sketch out my own ideas for the spaces being flipped before they revealed the final product at the end of the show, but I would always have to throw in a cough or snuffle when my parents came in the room. However, now I am certain those fake coughs weren't fooling anyone.

My mom had caught on to my love for HGTV and indulged me by often being the guinea pig in which I would showcase my design creations.

When I was ten, my dad asked me where in the world I would want to go if I could go anywhere. I had recently been given a Frank Lloyd Wright coloring book and managed to color every house within a matter of days. I was fascinated by the obscure shapes and spaces Mr. Wright was able to create. Because of this, it didn't take long to answer my dad's question. "Chicago!" I said proudly and with enthusiasm, knowing that Mr. Wright spent his early career dazzling the city with his masterpieces. So later that year, my dad and I took a five-day trip to the Windy City. We visited a number of Mr. Wright's properties, including the Robie House and the Frank Lloyd Wright Historic District in Oak Park. The week was full of fascinating real estate which also included staying at the historic Drake Hotel and tours of the Willis Tower Tribune Building and the John Hancock Tower.

From then, my interest continued to grow throughout high school, and I was able to take my first architecture class senior year. I spent my entire high school career as a varsity student athlete but once I arrived on the hilltop freshman fall not planning to play a sport, I found myself with ample free time. I decided to use that as an opportunity to further explore my interest in the real estate industry. I met with real estate professors and tried to attend local real estate events in D.C. to learn as much as I could. I became involved in both the undergraduate real estate club and the Public Real Estate Fund, which actively manages a $200,000 fund investing in both publicly traded REIT's and CMBS.

During the summer after my freshman year, I gained hands on experience working for a general contractor back in Rhode Island. Working on site, I learned how to properly manage a project schedule and budget, which translated well into my internship with Gilbane Development in Washington, D.C. over the summer of 2017. Known as a national developer with expertise in multifamily and private student housing projects, Gilbane provided me first hand exposure to all aspects of the development process. I am extremely thankful for the opportunities they provided me to work on a number of different projects, including a RFP in which Gilbane was subsequently shortlisted, and developing a new template to analyze comparative rents used in underwriting all new projects.

My time spent at Gilbane confirmed further my passion for real estate and interest in development. The mentors I gained through that experience have encouraged me to continue to pursue my increasing interest in real estate through this book.

Without over simplifying the industry, one thing I have learned about real estate is that it is driven primarily by the cycles of supply and demand. Developers want to operate in markets where demand is high and historically, those markets have been along the coast where prices are at a premium given the close proximity to the water. However, with today's understanding of climate change and rising sea levels, this behavior may soon be seen as irrational given the increasing risks and costs associated with developing in coastal markets.

Growing up in Rhode Island, also known as the "Ocean State" because of its over 400 miles of coastline despite its small size, I certainly understand first hand Mr. Nichols' blue premium theory that I mentioned earlier.

I spent every waking moment as a child in the summer near or on the water.

There was a small clam bed near the neighborhood I lived in and I would often look at the mornings newspaper to check when low tide was. I would then bike down to the beach and wade neck deep into the water and begin digging with my feet. Rhode Island is known for its clams, quahogs, steamers and littlenecks and that is why the popular comedy cartoon *Family Guy* is based out of the fictional town of Quahog, RI. The clams near me thrived in very sandy locations so whenever I wasn't in the water, you could find me building sandcastles. Long after my siblings and friends would wrap up their day at the beach, you could find me continuing to build up until the tide came to wash it all away. Going to the beach provided me with an endless canvas in which I could experiment with developing my castle/neighborhood/or masterplan ideas.

With my personal connection to the ocean—and my fascination for the supply and demand nature of the real estate industry—I have become increasingly interested in the industry's obsession with water front properties. It is this interest, in light of this past hurricane season, that has led me to try and figure out what's going on and perhaps venture a guess as to what we can do about it.

Throughout my research, it became increasingly apparent that I would not be able to reflect on how all types of natural disasters impact different real estate markets. While a significant portion of my motivation for initially taking on this task stemmed from my reaction to the 2017 hurricane season and extreme flooding, over the past several months the U.S. has been hit with a number of other devastating natural disasters including the California firestorms and mudslides, tornado outbreaks in the central and southeast U.S., and a "Bomb Cyclogenesis" along the East Coast. The majority of this book will focus on coastal and flooding natural disasters and will not explicitly address these other types. Furthermore, I will not address in detail *every* market that is at risk for coastal flooding, but many of the findings and suggestions discussed in this book can be applied to those markets not mentioned.

Admittedly, I'm no expert on real estate, natural disasters, or the relationship between them—there are people with years and years of experience in each of these fields, and their overlap.

But don't let this stop you from reading on. Like you, I'm curious to turn over every rock and talk to anyone I can to uncover what's *really* going on and what we *really* can do about it.

In writing this book, I aimed to educate myself on the implications natural disasters have on the real estate industry, and I am sharing what I've learned from speaking with some of the world's leaders on the subjects to be covered. Like many of you, I'm sure, I was bombarded with news articles during the 2017 hurricane season, but I felt there was a lacking element to each story. The majority of them were centered around eye-catching headlines or statistics detailing the severity of the season. Some of them addressed the timeline of recovery and restoration efforts, but very few adequately addressed the need for more proactive, risk mitigating efforts.

While a goal of directly influencing the way in which people live and develop along the coast may be out of the scope of this project, in writing this book, I hope to draw more attention to the rising risks natural disasters pose for coastal real estate markets and help progress the conversation on how we can develop smarter along coasts moving forward. My intention in writing this book is not to condemn the real estate community. I have a genuine love for real estate, and I hope to one day have a career in the industry. My intention is to better understand the mechanisms currently in place that have allowed large damages to accumulate and hopefully discover a path forward that will allow development to prosper, while also considering the risks and minimizing the costs and damages families deal with following a natural disaster.

After reading this book, you will have heard from experts from a variety of real estate backgrounds including property managers, private developers, fund managers, the Chief economist for the National Association of Realtors, as well as members of FEMA's disaster response unit, the National Flood Insurance Program, private natural disaster insurers, and leading non-profit organizations attempting to catalyze a change in our approach to coastal natural disasters. After analyzing the issue through these different perspectives, you will be well equipped to help bring awareness to this issue and begin to curate your own solutions to the problem. If there is one thing I have learned through this process, it is that a united, group effort is needed in order to accomplish tangible results and preserve the future of our coastal real estate communities.

My hope is for this book to reach not only real estate professionals with current exposure to coastal markets, but also the millions of victims of coastal flooding and others that currently reside in coastal communities but have not yet been burdened by a storm. I hope to clarify for these victims any aspect of their experience that may be unclear, whether it deals with insurance or the disaster relief packages provided by the government. Finally, I hope this book can reach those in political power, whether that is federal or local government, and light a fire under them to put aside the politics and start proactively addressing the vulnerable state of some of our country's most cherished coastal cities.

CHAPTER 1

THE EXTENT OF THE PROBLEM

―

Now that you know a little more of my personal story and motivation for picking this topic, let's now identify how bad the problem really is and why it will only continue to get worse if we don't start taking the appropriate steps to fix it.

In the introduction, I talked about Nichols' 'blue premium' theory which can help explain our fascination with coastal real estate. There has and continues to be significant over development along coastal communities. People continue to bet heavy on risky assets, but at what cost?

Well, according to a Risky Business national report on climate assessment, the average cost of a coastal storm will increase from $1.5B to $3.5B over the next 15 years.

In the chart below, taken from the National Oceanic and Atmospheric Administration, it is clear that the cost of damages continues to rise. This is due to the combination of an increase in development along coastal communities and the increasing severity of storms.

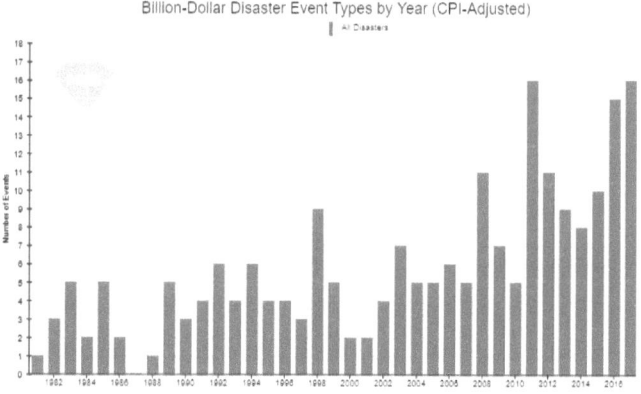

As an industry, we need to consider this problem and become smarter about developing in coastal communities.

MOVING BEYOND RESTORATION

Historically, a significant portion of the costs of damages have been covered by federal government aid packages. After a long and complicated process, Congress approved $56 billion in

Aid for Hurricane Sandy victims in 2013. This was in addition to a $9.7 billion increase in FEMA's borrowing power in 2012. Aid following Hurricane Katrina amounted to $114.5 billion. Immediately following Hurricane Harvey, the senate signed a package for $15.3 billion, followed by another package in October of 2017 for $36.5 billion, towards both Texas and Puerto Rico. We are spending billions of dollars every couple of years simply to restore to pre-disaster conditions, only to be sitting ducks for the next inevitable storm.

People need to understand that funding these costs through aid packages and simply restoring markets is not a sustainable approach to the problem. After all the money spent on Sandy, New York is still extremely vulnerable to coastal storms. People must begin to acknowledge that we need to go beyond restoring markets and brainstorm ways to build smarter and reform building codes and insurance distributions to reduce the impact of natural disasters moving forward. We need to build to sustain future storms, putting pressure on local jurisdictions and developers to bear the cost of sustainable development, instead of placing the aid burden on tax payers around the country.

CHANGING OUR SHORT-TERM MEMORIES

Celeste Hammond, a professor of law and director of the Center for Real Estate Law at The John Marshall Law School, said during an interview in 2018 with Bisnow, a commercial and residential real estate news outlet, "Increasingly, investors need to do due diligence on concurrent environmental problems and climate change threats to real estate they are using, buying or developing. It's a whole new world."

Currently, public real estate investment trusts (REITs) are not fully considering the implications of changing coastal weather. According to a recent report from BDO, an accounting and consultant firm, 97% of public REITs identified natural disasters as a business concern. However, none of them have acted on that concern and left a specific area due to a natural disaster according to Stuart Eisenberg, a BDO partner and national leader of real estate.

Given the blue premium, profitability for real estate investors is strong in coastal communities and industry professionals soon forget the devastating destruction storms can cause.

Michele Sansone, XL Catlin President of Property and Engineering, said in another Bisnow interview, "History has shown that we really have short memories. There will be a few that get fed up and want to stay out of those markets. Others will see it as an opportunity to pick up on some reasonably priced real estate."

STORMS ARE NOT SLOWING DOWN

While there is clearly an increasing problem along coastal communities without even considering climate change, the reality is that rising sea levels and increasing storm surges are only compounding the already apparent problem. There has been no shortage of news articles drawing attention to the increasing risk of future hurricanes. According to a New York Times article published in the fall of 2017, ten of the fifteen most active hurricane seasons since Antebellum America have occurred in the past two decades.

While there are differences in final numbers among sources, there is consensus that the costliest hurricane season in U.S. history was in 2005, which included Katrina and three other major hurricanes and cost between $140-$159B in total damages. However, experts agree that 2017 has become the new record, with most estimates well above $200B. Out of 17 storms in the 2017 season, four of the seven hurricanes reached category 3 or higher. These included Harvey, Irma, Jose, and Maria.

Jeff Masters, a meteorologist of Weather Underground stated in a National Geographic article that "The brutal 2017 season was an awful reminder of the huge hurricane vulnerability problem we face, and how underprepared we are for a potential future where the strongest storms get stronger and push their storm surges inland on top of steadily rising sea levels."

Furthermore, scientists from the National Academy of Sciences published in the fall of 2017 that a once in 500-year flood may occur once every 5 years by 2030.

There is a common misconception that this phrasing means a once in 500-year storm will only happen once every 500 years. So, if we just had one, then we aren't due for another one for another 500 years. This is incorrect. Instead, a once in 500-year storm means that there is a 1/500 or .2% chance that this storm will occur each year. That means that the likelihood of a similar storm happening next year does not change simply because one happened this year.

In coming to this daunting conclusion, scientists studied flooding from 850 C.E. to present day along New Jersey's coastline.

The findings included that "Until 1800, flooding of 2.25 meters above sea levels, took place on average once every 500 years. From 1970 to 2005, floods of that same height had a probability of occurring once every 25 years. To project that trend forward, the team then used models recently developed to analyze Antarctic ice sheet collapse. When combined with projected sea level rise, flooding of 2.25 meters-enough to do tens of billions of dollars of damage-could take place every 5 years from 2030 to 2045."

With this increasing trend of rising sea levels and flooding risks, we need to address the vulnerable state of our country now more than ever.

The risks and problems associated with coastal communities goes beyond hurricane seasons. In the process of editing this book, Boston has been hit numerous times with devastating nor'easters in early January and March of 2018 that have caused serious flooding. In addition to the frigid temperatures, these nor'easters brought large storm surges causing significant flooding in East Boston and specifically, the Seaport District.

Boston Mayor Marty Walsh told NECN in an interview, "As we saw with the coastal flooding in the Seaport in January, there is a need to proactively plan for our changing climate, which is why we're integrating climate resilience into all aspects of city planning moving forward, beginning with a Climate Ready project in South Boston."

Later on, in Chapter Eight, we will explore in further detail the unique risks associated with coastal real estate in Boston. New York is also a city that was impacted by these storms and is one we will also talk extensively about in Chapter Seven.

Even five years removed from Super Storm Sandy, on the Anniversary of that natural disaster, researchers from Climate Central determined that New York is the most vulnerable U.S. city to future storm surges and sea level rises, "with 426,000 people living on land that is imperiled through 2050."

After Sandy cost nearly $75 billion in damages, ranking itself in the top five costliest U.S. storms of all time, the city of New York still has plenty of work to do to mitigate future risks. In Chapter Seven, we will address some of the things they have already done in that regard. With over a trillion dollars in real estate assets that are sitting ducks for the next natural disaster, I set out on the mission to learn from and report on the experts already fighting this problem in hopes to preserve a prosperous future for our beloved coastal real estate communities.

PART II

HOW DID IT GET THIS BAD?

CHAPTER 2

THE FLOOD INSURANCE PROGRAM

"The Federal Government is not an insurance company. There is no other insurance company that I know of that outsources literally everything that they do." John Doe, a high-ranking official at a subcontractor for the National Flood Insurance Program

I am sure you understand all too well at this point that the purpose of this book is to analyze the impact natural disasters have on coastal real estate communities and to distinguish proactive measures we can take to mitigate risks moving forward. The analysis provided in this book, however, would be incomplete if we did not take a deep dive into the intricate workings of how flood insurance is currently provided.

(The identity, position, and company of the individual interviewed for this chapter have been concealed per their request given their position and affiliations with the government.)

Before speaking with Mr. Doe, I had very little knowledge about how the National Flood Insurance Program worked. I am sure, like many of you, my knowledge was limited to NPR or Reuters headlines such as "Flood Program Drowning in Debt" and "Harvey may add to debt woes of U.S. flood insurance program." I had read countless articles with pessimistic outlooks on the future of flood insurance. Before I could analyze the economic state of the program and find ways to stabilize it, it was imperative that I first obtain a solid understanding of how it all worked. It was through my conversation with Mr. Doe that I was able to move beyond the headlines and understand exactly what kind of monster we were dealing with.

A PRIMER ON THE NFIP

The Federal Flood Insurance Program works very similarly to McDonalds.

That's right. They aren't flipping your favorite burgers at three in the morning, but a large, franchised fast food company is a great place to start when understanding the NFIP.

The NFIP consists of 70 or so insurance companies, most of which you are probably already familiar with and have their jingle engraved into the back of your mind. These 70 or so companies write policies on behalf of the federal government. Each of these companies earn a small commission for the work they take on. Though the companies pay the claim initially, they eventually get reimbursed by the government, not only for the damage and loss expenses, but also for the claim handling process expenses.

Very simply, the federal government outsources the vast majority of day to day operations of the NFIP to these different insurance companies.

So where does the subcontractor group fit into this equation?

Well, the subcontractor group is a contractor to the NFIP, which is managed by the Federal Emergency Management Agency (FEMA) within the Department of Homeland Security. I'll get into what their current role really means, but it is interesting to know that in its initial scope, the subcontractor group was contracted to set up and handle the claim and underwriting operations for the NFIP. However, they won another contract that allowed them to migrate all of the FEMA systems under one roof, which the subcontractor group now manages.

In the words of Mr. Doe, "the subcontractor is in charge of all the systems related to the National Flood Insurance Program, and because they have all of the systems, they can see which of the outdated systems should be upgraded or replaced. They give the NFIP a lot of consulting advice around their systems and core processes."

What this all really means is that, at the end of every month, all of these insurance companies need to report back to the government the policies they have added or subtracted in the prior month, the premiums they've collected, and the claims they paid out. What generally happens with 70 insurance companies is that there are 70 different systems to compile this type of data.

Not extremely efficient . . .

So, the subcontractor basically maintains the federal government's official record. At a certain date every month, the subcontractor collects all this information from the underlying insurance companies. "The subcontractor cleans it up and makes sure it's all ready to go," says Mr. Doe. "And then the subcontractor uploads it into the federal mainframe which is a pretty old computer system." This system then spits out all kinds of reports related to the operations, including the finances, policy growth, and claim handling process of the program that the subcontractor and FEMA can then analyze.

Put another way, at this point the subcontractor is responsible for making sure all the data gets properly digested.

Looking forward, however, their long-term goal is to modernize the mainframe system. "The subcontractor's role right now is maintaining a dinosaur that the government has been using for decades and providing advice on how to modernize this process." In addition to getting the system up and running, the subcontractor has an oversight role of the underwriting and claims the handling process. In this process, they audit a number of claims and policy transactions every month to make sure the insurance companies all handle the claims properly.

If a customer files an appeal, "the subcontractor will handle those situations directly," says Mr. Doe. That process would look something like this: an appeal is received. The subcontractor then addresses the customers complaint by looking at the company and how they handled the claim. Then, they report their findings to the federal government, which ultimately decides whether to overturn or stand by the initial decision the insurance company made.

SO, WHAT DETERMINES WHO IS IN THE PROGRAM?

FEMA has flood plain maps that are used to ultimately decide if you are required by law to have flood insurance. If you live within the 100 year flood zone determined by FEMA, you are required to have flood insurance, but only if you are paying a mortgage on your house. This is because mortgages are federally insured.

The federal government basically says that if you are going to buy a mortgage that is federally backed, that is also in a flood zone, then you must also purchase flood insurance. While the preference is for everyone to have a policy, this will likely never happen and so this mandate helps to ensure more people will be covered than if it wasn't required.

While this mandate may appear to be an effective policy that works to get more people insured, through listening to Mr. Doe, I quickly became aware of its underlying inefficiencies.

The first inefficiency is that the government places the onus on the real estate and mortgage companies associated with the purchase of a property to ensure a flood policy is provided. The government basically tells these companies that if there is a transaction of property, and a mortgage is being placed on the house, then they must ensure the customer is purchasing flood insurance. There are standardized paperwork and audits for these companies to ensure that this occurs. "Up front, when a home is sold or resold in one of these areas, there is a lot of emphasis for the mortgage and real estate companies to 'check off the box' for flood insurance. They must initially make sure you have a flood policy, but beyond that, their role suddenly becomes less clear," says Mr. Doe.

What Mr. Doe means by this is that upfront, there is a lot of oversight; the mortgage company manages the mortgage, the insurance company manages the insurance policy, and FEMA is overseeing the flood aspect of it all but not the mortgage. This means that even after a couple of years go by, there is not that same collaboration of oversight that was present when the policy was first established. The oversight becomes extremely less clear because the mortgage companies don't have direct access to the FEMA data base with flood maps that detail who does and does not currently have a policy.

So, if someone cancels their policy or stops making payments, there is not an effective system in place that goes back to the customer and says, "hey, you are still making your mortgage payments but you let your insurance lapse, so you need to reinstate your policy." One of the reasons why FEMA cannot effectively oversee this process is because there is nothing in FEMA that tells them when all of their customers pay off their mortgages. "That's simply not information that is recorded anywhere," says Mr. Doe. "The subcontractor doesn't know after that initial oversight who does or does not still have a mortgage and the subcontractor is not following up with their insurance payments."

There is clearly room for improvement.

With all the technology these days being used by large companies to aggregate big data on customer profiles, it seems ridiculous the government cannot find a way to keep track of this information effectively. The onus is clearly not on the insurance agencies contracted out by the government to follow up on payments because, after their commission on sale is recorded, the rest of their expenses are reimbursed by the government. Therefore, the impetus is on the government to update this system and come up with an effective way to track who is required to pay.

While the lack of effective policy and information tracking certainly hurts the federal program's bottom line, the manner in which they underwrite properties is economically unsustainable.

The NFIP is a government program and is not incentivized to be profitable. This means that no private carriers wish to compete in flood prone areas because everyone will already be receiving subsidized policies from the government. "The price for federal flood insurance is not a market set price," says Mr. Doe. "Even in flood zones, the pricing formula is not complicated; it's actually very simplistic: you choose a limit and a deductible based on the certain zone and size of the property." The subsidized policies the government provides are therefore not reflective of the potential risks associated with these properties and therefore certainly don't correlate with the potential claims to be paid out. These are both inherent attributes of any well-run insurance company and can explain why the whole federal program is bankrupt.

A typical private insurance company is going to try to squeeze the margins as much as they can and still be profitable. This means they will charge an amount for policies that pays for claims and operating expenses while also leaving a little profit left over. But that is not what the federal government does.

"It operates at a deficit at all times," says Mr. Doe. "The premiums FEMA has been collecting aren't even enough to service the debt on the deficit. That's how bad this situation has become."

Hearing that remark caused me to pause for a second to remember an article that I am sure many of you have seen that was published only days prior to my interview with Mr. Doe. It highlighted how the program was already in the hole $26 billion following Katrina, Sandy, and a few other storms. In light of the fact that the program could not even service its own debt, let alone provide the necessary funding for Harvey and future storms, the government decided to forgive $16 billion of that amount.

In response to this statistic, Mr. Doe said, "If the angle of your book is to identify inefficiencies, there are plenty to pick from. One of the questions you have to ask is if you can even squeeze out enough efficiencies to save the program."

Even if we are able to address a number of the easily identifiable inefficiencies with the program in its current form, it's unlikely the program will ever be sustainable as a government subsidized insurance provider.

Given this, there has been a lot of talk recently about the possibilities of privatizing the practice. Doe explained that the program is enacted by Congress and it's refreshed or renewed every three to five years. Our current program was supposed to expire on September 30th. In order for it to be renewed, there must be legislation saying they are calling for an extension of the flood program for x number of years. It's during these renewal process periods that changes to the program are enacted. On September 30th, it was clear the government was not in a place to properly address everything that needed to be fixed and decided to push out the decision until December 2017.

The main point, Mr. Doe says, is "that when the program is ultimately renewed in December, there is a huge expectation that this time, the federal government will change the program to allow private carriers to replace the government in certain flood prone geographies." Reform is desperately needed and there is currently a lot of interest in making the program more efficient and in allowing private companies to play ball.

When asked what a shift towards private insurers would mean for his business, Mr. Doe believed the change would have a long tail in having any real impact to his company's role. Whether the government is selling policies through seventy or seven companies in the near future, the same work must still be done in the immediate future. Even when private insurance companies begin to compete with one another, the rates they will be charging will still be a significant multiple higher than what they currently pay.

Because of this, Mr. Doe believes the government will intervene in certain situations and declare that the price increases for some is too much to bear. In that case, he feels it's likely that the government will decide to keep those policies knowing they will incur a loss every year. "I still think we would be left maintaining government subsidized policies and claims, but they will likely be focused on the poorer communities right in the middle of flood zones."

CHAPTER 3

UNDERSTANDING THE EXPERTS

———

"99% of people don't take [climate change] into consideration on the private real estate side." Aakash Bhatia, ASB Capital.

Given the short-term dynamic in commercial real estate, it is extremely unlikely the industry will slow down the development along coastal communities. I'm sure this comes as no surprise to some of you. If the demand is there, why do anything different?

But in order to understand fully the private developer's role in this issue and potential role in its solution, I reached out to the experts themselves to understand how they are currently approaching the topic of climate change and natural disasters.

While a majority of my research and interviews have been focused on those involved in government agencies or sustainability programs centered on creating protective solutions for coastal real estate markets, I wanted to understand first hand how the private side of the industry is addressing the imposing threats of climate change. So, I looked to Mr. Bhatia for help.

Aakash Bhatia is a fellow Hoya, a recent graduate of the McDonough School of Business, and he is now an investment analyst at ASB Capital Management, one of the largest institutional investment management firms in the D.C. area.

ASB is currently exposed to a number of coastal markets vulnerable to the future impacts of climate change, including a number of properties in downtown Manhattan, Miami, and two office properties in Boston's Seaport area. Bhatia mentioned that he has experience underwriting deals in Houston, but, currently, their firm does not manage any Houston assets.

In reflecting on his own experience in these markets, Bhatia said, "Boston does a decent job in terms of regulations focused on the impact of climate change. They have a Climate Ready Boston report which is very extensive. And we have had facade inspections every five years for our office properties in the Seaport area. New York can do a better job, though, given the amount of low lying land so close to the river. New York is tricky because, historically, it hasn't had many floods."

Both New York and Boston are vulnerable markets that deserve plenty of attention, but nearly 1,500 miles down the coastline is yet another market that carries its own risk: Miami.

"I actually just got off the phone today with people talking about a property on the beach in Miami," says Erik Rutter from Tishman Speyer.

In addition to Bhatia, I also reached out to Rutter to gain yet another perspective from the private real estate industry. Mr. Rutter is positioned in one of Tishman Speyer's downtown Manhattan offices and has extensive market knowledge in that area as well as Miami.

"There is a large trend right now to develop new, ground up office buildings with curtain wall facades," said Mr. Rutter.

In the real estate community, curtain wall simply refers to an enclosure system made of glass and sometimes steal that performs against wind loads and seismic activity, but is not part of the superstructure of the building. Think of any buildings you have seen in Manhattan or another big city that appears to be all glass windows on the side, and you've got yourself a curtain wall. "In Miami, however, you see a lot more concrete being used because the curtain walls do not hold up as well during hurricane seasons as concrete based facades that are part of the building's superstructure," Mr. Rutter notes.

"So maybe, we thought, there is an opportunity to deliver a curtain wall product in that area, an area that usually doesn't have curtain walls. Therefore, we would be able to charge a rent premium because we would then be offering a product in short supply in that market."

With regards to the research I have done, I was relieved to hear the local government is taking initiative and limiting the use of curtain walls along the beach. Mr. Rutter, however, raised an interesting point that for some developers, in markets where the government might not take a similar stance, delivering a unique product to the market would justify the risk of having a more vulnerable asset to climate change and future storms because of the rent premiums to be realized.

From this, it's easy to see how industry professionals are not incentivized to develop sustainably and responsibly, but, instead, are incentivized to build riskier assets by the large tenant demand for ocean front properties and their willingness to pay the "blue premium."

Mr. Rutter acknowledges there are gains to be realized by providing a desired product, such as curtain walls, in a market that is undersupplied, like Miami due to regulations; in a market like Manhattan, though, curtain walls are more of a requirement. In Manhattan, where the government does not take a similar stance as in Miami despite it still being an extremely vulnerable market, you must deliver a curtain wall even if it is more susceptible to damage from hurricanes because that is where the market is. If you do not, you would risk producing below market rents and your investment would bust.

BEYOND INSURANCE, THERE IS LITTLE CONSIDERATION

When asked how ASB currently addresses the concern of climate change to their portfolio, Bhatia pointed to a single line item on their projected income statements: insurance.

"People think about insurance a lot more when talking about assets in risky markets like these. The thought process starts early in the course of underwriting the property. You just want make sure you're covered." When assets are in vulnerable locations on the asset management side, the insurance associated with this risk is something you certainly need to budget for every year within your pro-formas.

Mr. Rutter also gave his opinion on the matter and said, "If you are building on or near the water, it is certainly going to be on your mind. If you are building in an area that is hit all the time like hotels in South Beach Miami, it's going to be top of mind. But, they already have a ton of precautions in place that are government mandated that we follow."

Apart from insurance and the current government mandates, such as limiting curtain walls in Miami, Bhatia notes that it's not a huge component of the underwriting process; it's simply something you keep in mind and adequately budget for. Thinking about climate change and spending more to mitigate risk beyond what is covered in your insurance line item "is not something a lot of people think about in downtown Manhattan. 99% of people don't take that into consideration on the private real estate side."

Bhatia confirmed my own suspicion that people in the private real estate market don't think more about this topic, because if they did, I'm sure they would have taken further risk mitigating actions up to this point and we wouldn't be left with the catastrophic problem we have today. But I really wanted to dig deeper into this issue and understand why. Bhatia had this to say:

"When you are building out your own pro-forma and competing against others to bid on or purchase a property, if one investor budgets $500k in Capex to fortify a property against a storm and flooding, then they are going to be at a HUGE disadvantage in the bidding process because naturally they cannot bid as high as the others because they have this in their pro-forma. So, until everyone gets into the same mindset with underwriting climate change risk, then I think its going to be tough to expect anything to happen."

THE FOCUS IS NOT ON FLOODING

While it may be easy to understand that the private developers are not incentivized to invest in this cause beyond budgeting for their insurance, what might surprise you is that flood insurance is not even the biggest concern for asset managers in coastal, low-lying markets, especially New York.

In fact, specifically in New York, the market places a premium on terror attack insurance following the events on 9/11. There are plenty of markets that don't even consider insurance like this, but, in downtown Manhattan, terror attack insurance is thought of as being far more important and risky than flood or other damage insurance with regards to climate change. "In a market like New York, the terror insurance is often a sizable multiple larger than that of flood," says Bhatia.

Despite a recent Climate Central report published in the fall of 2017 that ranked New York City as the most vulnerable city to major coastal floods, climate change is not top of mind for the individuals involved in this market. The closer you get to Lower Manhattan, the more expensive your insurance becomes, but Mr. Rutter further confirms what Mr. Bhatia said, that "your terrorism insurance will most likely be higher than your natural disaster insurance."

To the extent that anyone is willing to mitigate risk above and beyond the government mandated precautions is a question of whether or not you see a return on those costs. "You're not going to increase your construction costs for no reason. There has to be a way to net out that additional spending," says Mr. Rutter.

Specific to the Lower Manhattan market, if you increase security in a highly terror vulnerable area, people like that. Take the Freedom Tower, for example. I recently watched a documentary on the construction of the building, and it has easily become one of my favorite structures in the world because of the thought and detail that went into its design. For good reason, the Freedom Tower has higher terror insurance than a property on Park Avenue even after it took extreme risk mitigating action to reduce vulnerability to terror attacks.

"The Freedom Tower was government mandated to have an extreme amount of security," Mr. Rutter notes. The entire bottom ten floors of the structure were made out of thick titanium and concrete, something you could never drive a car into. Using the Freedom Tower as an example of mitigating risks and referring to mitigating risks associated with climate change, Mr Rutter said, "I just don't see the impetus on the developer, unless it doesn't cost any more money to mitigate, but it does. The impetus is likely on the government, similar to the Freedom Tower."

When asking Mr. Rutter about the impact Sandy had on the

Lower Manhattan commercial real estate market, he had this to say: "When you step back and see the world of hurt the downtown market was in following 9/11 and before Sandy, it was slowly climbing its way back. I don't think Sandy destroyed the market the same way that 9/11 did; it felt more like a hiccup in the wake of both 9/11 and the financial crisis of '08-'09."

WHO HAS THE IMPETUS THEN?

This response left me a little disheartened that the industry I am so eager to join, and hopefully one day have a career in, is neither taking this issue seriously nor properly underwriting the process.

But, I now understand why. There are no incentives or further regulations in place to properly incentivize and change the private industries' course of behavior. They will continue to supply what the market demands. Bhatia believes part of this is because there is no return on investment for the private side to further mitigate risk.

While the development community has yet to respond actively to climate change initiatives on a large scale, Mr. Rutter draws our attention to the interesting ways in which the development community has responded to Leadership in Energy and Environmental Design (LEED) certification initiatives.

LEED certifications are now a big trend in the design and construction process. Mr. Rutter notes that there are a lot of foreign investors in U.S. markets, particularly in Manhattan, that promote LEED initiatives. The Norwegian pension fund is one of the largest deployers of capital into the U.S real estate market, and they care deeply about sustainability. "If you are working with them, they require you get your building certified." This is because the impetus is with the investors given the payback period associated with these types of extra capital expenses; you see a lot of developers try and get LEED certified.

If you have a horribly outdated building with inefficient equipment, there is a big capital expense associated with restoring this equipment. But, once you do, that equipment has a payback period in which it will start saving you money through more efficient energy and water use. "When you perform improvements under LEED," Bhatia notes, "you aren't just putting money in. You are able to see a tangible return on your investment through reduced energy bills, water usage, etc."

For flood mitigating actions, there has been no similar ROI or incentive in the past for being proactive with regards to climate change. Therefore, this means that, at the moment, there is not similar support from real estate equity investors as there is with LEED initiatives. The reason for this has to do with finding out how much equity investors care about weather proofing their properties beyond what is government mandated and regulated. Mr. Rutter believes that "there are not a lot of investors out there who would say "go and spend the extra dollars to weather proof the property." This is because there is not the same anticipated payback period associated with capital expenses.

HOW ABOUT THE TENANT'S DEMANDS?

Another important aspect to consider in this equation is the tenant population you are trying to attract to your property. How much do they care if a property is weather proofed? How much do you care? In New York, it's clear a lot of the younger millennials care about sustainability. "You'll see a lot of tenants say they prefer to be in a sustainable building, but I have never heard a tenant say they care about being in a more weather proofed building." In order for a tenant to truly care, the property would have to be significantly safer than the other options in that market. If the tenants did take this into consideration, it may be possible to charge a rent premium to cover the added costs of mitigating, similar to some properties that display LEED certifications; but, according to Mr. Rutter, that is not where the current market stands.

THE GOVERNMENT'S IMPETUS

Both Bhatia and Rutter have clearly outlined why the private industry has not already become actively invested in this cause.

It's the lack of incentive on the investor and developer level that has allowed this problem to grow with no reins. If the private developers won't change their behavior, neither will the investors because the tenants will continue to demand their ocean front views and are willing to pay for them. As a result, we must look to the government to help catalyze a change in the industry's unsustainable behavior.

Mr. Rutter mentioned that a great example of when the government played an active and influential role in the changing dynamic of a real estate market was following the attack of 9/11 when the lower Manhattan market was destroyed.

"Prices began to fall significantly, and vacancy skyrocketed because people were saying Lower Manhattan was unsafe due to one terrorist attack," says Mr. Rutter. In the current discussion of climate change, this reaction was unknowingly a proactive reaction. People began to retreat from Lower Manhattan, one of the most vulnerable sub markets within the NY/NJ area.

At this point in our discussion, all I could think was that this reduced interest in real estate along the river's edge due to the terror attack caused people to retreat and actually began reducing the future risks and damages associated with climate change and rising sea levels in the area.

However, in fear of a further market collapse to one of the most highly valued real estate markets in the world, Mr. Rutter recalls that the government "responded by handing out incentives to buyers, developers, and tenants to relocate their firms to Lower Manhattan." If they moved, they received bonuses that effectively allowed them to avoid paying taxes for a long period of time. When the Port Authority moved to the new World Trade Center, "on paper their rent looked like $60 a foot, but with government incentives, they ended up paying around $30 a foot."

While the government's incentives surely re-introduced further damages associated with climate change by bringing people back into Lower Manhattan, their effectiveness in providing a catalyst for change impresses me. Now I am left thinking, what's stopping the government from finally realizing the extremely vulnerable state of some of the country's coastal communities and the projected damages that come along with them? And, in the same manner as demonstrated before, how do we incentivize them to move away, or up to higher ground?

PART III

ENTERING A PROACTIVE MINDSET

CHAPTER 4

GOOD, FAST, CHEAP

"You can find online that Einstein defined insanity as doing the same thing over and over again and expecting a different result." Ed Thomas, National Hazard Mitigation Association Volunteer

After seeing Mr. Thomas's name in a number of articles published following the slew of storms in 2017's hurricane season, and reading an article he co-published for the Brookings institute titled, "Reforming Federal Support for Risky Development," I knew I had to reach out to him personally. I encourage all of you to check out Mr. Thomas' article on your own time.

Early in his career, Mr. Thomas became involved with local governments through his work on housing and community development projects. This led to a career at the U.S.

Department of Housing and Urban Development (HUD), where he expanded his experience to include disaster assistance. This work focused on helping people get back into housing after some sort of natural disaster.

In 1979, the programs Mr. Thomas was involved in were transferred over to FEMA. After spending a couple years at FEMA as a federal coordinating officer focused on responses to disasters, he retired and shifted his focus to being an engineer. During that time, he worked on developing a process for executing safe and proper design before getting whacked by the next storm.

Over the past seven years, Mr. Thomas built upon his experience with natural disasters response by working as a volunteer for the Natural Hazard Mitigation Association (NHMA).

The NHMA is an entirely volunteer run educational charity that focuses on shifting the nation's approach to natural disaster response by advocating for and influencing policy reform.

By writing materials and educating at the grass roots level, the NMHA works to "change the way in which official U.S. policy has responded to natural disasters in the past, which has been to restore communities to pre-disaster conditions as quickly as possible. You can find online that Einstein defined insanity as doing the same thing over and over again while expecting a different result every time."

The NHMA focuses on changing the response to one that includes safe and proper design for disaster risk reduction. "We have a curriculum and provide people educational material so that they can go out and continue to advocate, which turns around common poor development practices and rewards good, risk reduction behavior," says Mr. Thomas.

In addition to their own educational publications provided on their website, the groundbreaking work done by this organization has frequently been published in U.S. World and News, the Houston Chronicle and numerous other media outlets.

I was extremely fascinated from the beginning to hear that Mr. Thomas began his work in the public works position with FEMA and has since transitioned over to the privately run NHMA advocacy and educational organization. I was eager to understand Mr. Thomas's observations and experiences between the two types of organization and their approach to and successes with this issue.

Mr. Thomas responded to this inquiry by first saying that some government agencies can draw the line and say, "we aren't going through this again. We are going to fix it and fix it right this time. But, those are certainly in the minority." This means that the vast majority of government organizations have an insufficiently short outlook on the situation, and their response lacks the adequate action needed to sustain stability within coastal communities in the future. He expanded further on this by explaining to me the Good, Fast, Cheap dynamic.

GOOD, FAST, CHEAP

Although a long-term approach is needed, the reason it hasn't happened is partly due to the fact that it goes against the human instinct of "I just want to get this over with and want to get back to the way things were as soon as possible." Mr. Thomas furthers this point by saying, "There is a triangle in play of good, fast, and cheap. If you can get a project to meet two of these, then you've done a good job. But, it's nearly impossible to hit all three."

This only makes the process and road ahead tougher.

If we want disaster response to be good and fast, it simply won't be cheap. Or, if we continue with the HUD and FEMA mantra of resorting to pre-disaster conditions, we will continue to have response be fast and cheap, but it won't be good.

Solutions to the issue of increasing risk associated with climate change in coastal communities must now be focused on breaking this tradition of cheap and fast responses.

Furthermore, Mr. Thomas notes that the days for blaming Mother Nature or using the excuse that an event was totally unforeseeable are well behind us. "Claiming that something is a 1,000 year storm and not doing anything to mitigate the risks because of this is balderdash," says Thomas.

Our focus should be on building more safely and not making these types of excuses like "we didn't see it coming." To the extent that we ever knew what a 1,000-year storm was in history, we certainly don't know now. There are plenty of current known-unknowns with regards to impacts climate change will have, and our design approaches must adapt to this. We must move beyond the age-old traditions of natural disaster management policy that simply focuses on restoration to one that proactively assesses and mitigates future risks.

One specific example of successfully implementing the education and information currently available to us in order to develop more sustainably came from one of Mr. Thomas's pupils.

RESULTS, RESULTS, RESULTS

Alan Greenberg, now a well-known American architect, once attended Harvard's Graduate School of Design, and, during his time there, he took a class instructed by Mr. Thomas.

While in class, Mr. Thomas remembers Mr. Greenberg's fascination with coastal development and the regulations set forth by FEMA's flood elevation maps. He was eager to learn more about how FEMA established these flood maps and wanted to apply this knowledge to a sub development project in Houston.

When asked what became of Mr. Greenberg's project, Mr. Thomas said, "I just heard recently from him that the 500 or so unit subdivision he developed did not suffer any damage in the wake of Harvey."

Mr. Thomas believes this can be attributed to research and preparation Mr. Greenberg undertook to go above and beyond the mandated zoning requirement. Through this process, Mr. Greenberg discovered that FEMA's flood elevation maps had plenty of limitations. The maps themselves are based on historical events, and they fail to take into consideration any relevant information on sea level rise, climate change or changes to the environment. These changes include over-paving in certain regions like Houston with non-permeable materials, which increase the likelihood of flooding.

After learning of all this, Mr. Greenberg decided to simply not trust the flood maps provided by FEMA and, in building his subdivision, he used an additional five feet of free board elevation above FEMA's established flood elevation. This extra five feet was more than enough to keep Mr. Greenberg's development out of Harvey's way.

In the coming months, Mr. Thomas hopes to write an article or develop a lecture series for the NHMA to highlight this project. Mr. Greenberg's efforts provide a prime example of how proper sustainably focused design and construction can reduce the economic impact natural disasters have on coast infrastructure and communities.

There is always plenty of mention in the news about who is getting what disaster relief, but "the folks in Alan Greenburg's subdivision are receiving the best kind of disaster relief, which is none, simply because of safe and proper design," says Mr. Thomas.

Given the excellent design and its ability to withstand Harvey, Mr. Greenberg's subdivision will undoubtedly be worth more post-Harvey than pre-Harvey. I reached out to Mr. Greenberg with the hopes to speak in further detail about this project but, unfortunately, I have yet to receive a response.

IS THIS SCALABLE?

While the true value of Mr. Greenberg's subdivision might be realized following Harvey, in order to make this approach and design practice applicable to the broader commercial real estate and development community, and to understand why this has not already been widely adopted in the industry, we must consider the extra costs associated with pursuing risk mitigating construction.

Mr. Thomas notes that retrofitting past mistakes can become really expensive, but the costs associated with mitigating risk during the design and construction process from the beginning can likely be recaptured through increased safety and deductions in insurance premiums. "Maybe with low income housing it can be tough to push for mitigating risk," Mr. Thomas notes, "but they are the ones that can least afford to be flooded, so special attention must be given to this demographic when establishing an effective approach to mitigating risk."

As a society, we have to make a decision:

Do we continue our practices of the past and continue to spend large amounts of money on disaster relief aid simply to restore and rebuild? Or, do we spend a little more money upfront now in order to protect our properties and reap the benefits of a stable and thriving coastal community?

As we have just seen from Mr. Bhatia and Mr. Rutter, the largest concern for developers is that they will pay a couple of percentages more in construction costs to mitigate risks associated with their properties and they won't get their money back.

However, Mr. Thomas thinks that safely designed houses will generate more positive attention and will allow developers to get that money back. Currently, the development community is in a unique position "in which they typically skate on a post disaster lawsuit because of the fact that typically developers finish a project and unload the asset." The costs associated with risk the developers take on is often externalized back to society, where federal taxes will help support the natural disaster victims.

Although reform involving developers and the actions they take would be ideal, Mr. Thomas says this is unlikely to occur because he feels "the courts will simply not pierce the corporate veil." This is because, while the local governments are at risk due to over waterfront development in their jurisdiction, they are also the beneficiaries of such events. For local governments, over development along the coast means a lot more tax revenue from properties being developed and congregated along the coast, which is again another reason why this issue has not been properly addressed yet.

Just as I was having an "AH-HA!" moment listening to Mr. Thomas' story of Mr. Greenberg, it became clear why actions like these have not become common practice and why it is so difficult to change our current behavior and response to natural disasters. The educational work the NHMA is doing is certainly leading the charge on sustainable coastal development. Their agenda of shifting the response policy from one of restoration to one of proactive mitigation effectively addresses the root of the problem we face today of increasing risks and damages along coastal communities.

Through Mr. Greenberg's development, we have seen the impact a proper, risk mitigating approach can have. He quickly addressed his distrust of FEMA's historical flood maps and begged us to ask whether we should do the same. The validation of these maps will be addressed later on, but, for now, we are left with the question: how do we scale the efforts and results exemplified by Mr. Greenberg?

CHAPTER 5

INSTITUTIONAL INTEREST

"The storm has opened up an interesting opportunity for us to try and drive these new initiatives." Andrew Green of Capital One, referring to their risk mitigating initiatives following Hurricane Harvey

In Chapter Three we heard from both Mr. Bhatia and Mr. Rutter about the influence institutional investors have on the development process. They introduced the concept of LEED certifications and explained the successful response the program has received from the development community. They briefly mentioned the interest institutional investors, such as the Norwegian Pension Fund, have taken to LEED initiatives and how this interest has helped transform the design and development aspect of the industry.

In order to understand this relationship further and determine if there are any key takeaways to learn from the LEED model, I connected with Andrew Green, Head of the Environmental Sustainability Office at Capital One. However, the process in which I was able to connect with Mr. Green was somewhat by luck and certainly more extensive than most of my other interviews.

Shortly after arriving to school in early September, I was sifting through the past few days of Bisnow articles sent to my email every morning, and I came across an ad for DC's third annual real estate sustainability summit. I began reading further about the event and became increasingly interested, so I pulled out an old visa gift card I had received years ago as a birthday present and spent the last of it on the $35 dollar ticket.

That Friday, I woke up at 7:30am, threw on my one grey suit and hopped in a cab off to the summit.

I didn't really understand what I was walking into.

I quickly came to the realization that I was easily the only person still in school and under the age of 27. The convention space was filled with companies tabling their products and new inventions, relating to virtually every aspect of the real estate. The day was filled with tabling events, panels on real estate sustainability and development, networking events and a keynote speaker that convinced the entire audience the future of construction is "prostruction."

His point was that buildings have been constructed using the same or similar materials for very long time, and currently there is a push for buildings to become more and more sustainable. He proposed that the most sustainable building would be one that is alive and grows on its own, patches itself on its own, breathes in the CO_2 and pollutants in our world and breathes out clean air. In his mind, "prostruction" was about changing the dead materials we build with and, instead, build with materials that are alive.

While it was a fascinating thought experiment at the time, and I am sure you, too, are now thinking how something like this could ever be possible, I gained the most from this summit through my interaction and follow up conversations with Andrew Green.

CORPORATE TAKES THE LEED

Historically, Mr. Green and his team's focus has been on making building operations more sustainable, including reducing greenhouse gas emission and energy consumption. Mr. Green highlighted a corporate sustainability policy in place at Capital One that all new buildings, either constructed or renovated for occupation, must be LEED Silver certified or above.

Mr. Green does acknowledge, like most other firms in the industry, that there are further opportunities for the firm to do more in terms of risk management and combating flooding and climate change, which they historically have not pursued. In the wake of Hurricane Harvey, these opportunities are now at the front of Mr. Green's desk; "the storm has opened up an interesting opportunity for us to try and drive these new initiatives."

One of these new initiatives is to focus further on the company's CDP report.

The CDP, also known as the Carbon Disclosure Project, is an initiative that focuses on tracking your greenhouse gas emission on an annual basis, setting reduction goals, and then publicly reporting on the progress of reaching those goals. It is something a lot of major companies, cities, institutions and universities do every year.

Many companies, including Capital One, do it at the request of large institutional shareholders. Mr. Green was proud to share that Capital One hit their second greenhouse gas goal four years early at the end of 2016, but he acknowledged that they will continue to set higher goals for that and other questions the CDP ask beyond emissions, such as how you incorporate climate change into your risk assessment and governance processes for your company; the bar gets higher every year.

This was the first time I learned of the CDP, but I think it is a fantastic initiative that Capital One and plenty of other organizations have undertaken to be proactive in their approach to sustainability and climate change.

What interested me most was that this project was initiated at the shareholder level. Mr. Green elaborated on this by mentioning that one of their institutional investors is the State of New York Pension Fund. Recently, this pension fund was involved in a proxy vote this summer with Exxon, another company they have invested in. In this proxy, the shareholders forced Exxon to start reporting annually on the exposure of their business to the carbon environment, and to start shifting towards a low carbon environment. For a company like Exxon, that's a pretty big deal.

In addition to this, the "State of California Pension Fund is another well-known institutional investor that has strong sustainability targets for the companies they invest in, especially around risk analysis. So, they request companies to do carbon disclosure reporting, as well," Mr. Green notes.

The fact that institutional investors see value in supporting sustainability efforts is a positive sign. Being able to drive sustainability efforts through institutional investors in order to catalyze a change in behavior could be one of a number of ways in which we are able to address the over development of real estate in vulnerable coastal markets.

If institutional investors begin to value the safety and longevity of real estate assets in vulnerable coastal markets and take responsibility for properly mitigating future risks of flood and other damages to a property, we may begin to see more resilient coastal communities.

Unfortunately, this is all theoretical. Institutional interest in LEED initiatives like reducing energy use and greenhouse gas emission may not translate to an investor interest in mitigating hurricane risks to coastal properties.

WHY IS THIS? TIME AND MONEY.

Two weeks prior to our follow up conversation, Mr. Green attended a conference in Boston that focused on sustainability improvements in the real estate environment. Specifically, the focus was on large development projects and situations in which institutional investors were involved. In situations like this, "it gets tough because of the type of investor you are dealing with."

Large REIT's and other institutional investors in real estate often buy up properties, lease them up, and sell them. In deals like this, the time horizon is often less than seven or ten years. With such a short investment period, there is little need to consider the long-term storm projections or risks associated with increasing coastal flooding. The payback period on costs associated with mitigating risk will likely be much longer than a REIT's typical hold period. This means there is very little incentive for them to consider this issue beyond what they are already budgeting for insurance.

It is unlikely that institutional investors will insist that the REIT's they invest in pursue sustainability efforts with regards to mitigating flood risk. Further capital expenditures and no realized return on that investment within the specified hold period would only hurt investor returns.

With the way the current industry works, it simply is not in the best interest of the institutional investors and the REIT operators to take this issue into consideration. The reality is that because of this, developers will continue to develop along coastal markets without aggressively considering the costs associated with a large hurricane.

Think of this as a mash-up of musical chairs and hot potato but with commercial real estate. Everyone is passing around large assets without putting in the necessary Capex to make it a safe and reliable structure to withstand storms in a longer hold period. At the end of the day, when the music stops, aka a storm hits, and the damages exceed what the insurance policy is underwritten for, who ends up paying for the risks all these developers took and didn't pay for? Tax payers, all through disaster recovery aid packages.

So how, then, is LEED different? Why have some institutional investors supported it and not flood risk mitigating initiatives?

As we already know, LEED stands for Leadership in Energy and Environmental Design. It's an international green building certification system that aims to reduce construction waste by using recyclable materials, and reduce water, energy use, and storm water run-off.

Depending on the extent to which your design and building process meet the standards associated with these initiatives, your building will receive one of four certifications: LEED, silver, gold, and platinum. This certification process has been extremely successful in shifting the design and construction focus to take a more sustainable approach.

For the most part, companies often pursue LEED certification only when it makes operational sense. As a large bank, Capital One is able to invest more in LEED because the excess capital needed to fund these initiatives is readily available.

"We want to make decisions with operation costs in mind," said Mr. Green, "and we choose LEED projects with good payback periods for each building's hold strategy."

Capital One owns about half of their corporate office and leases the other half. Among those they own, including an 8k person campus in Richmond, VA, a 7k person campus in Plano, TX, and their HQ in Tysons, VA, which will hold 6k people when it is completed, they "have a longer-term outlook and are depreciating the properties over 40 years or so." Given this longer-term hold period, they are able to invest in more LEED initiatives because the reduction in maintenance and operational costs of the building associated with these initiatives can be realized over this longer horizon.

Mr. Green mentioned that this is specific to their long term hold strategy and "a slightly different mindset is needed for a design, construct, and deliver manager who might not have a long term hold period," and is likely unable to be as aggressive with regards to LEED initiatives.

Up until now, there hasn't been any similar type of ROI for developers to mitigate their properties from flooding; "it's a pure risk play and that's about it." What Mr. Green means by this is that developers are currently not incentivized enough to invest in risk mitigating initiatives because there is no similar return on operational costs.

And, given the likely shorter-term strategies of most developers, it does not make financial sense for them to invest more in mitigating risk beyond their current insurance line item.

In an attempt to receive a return on this added Capex, "depending on how you structure your leases, you could try and pass the cost through to the client." However, current tenants and market conditions don't really value reduced vulnerability in the same way they value energy efficient and air quality in order to pay a premium for it.

It's for this reason that LEED is also a social responsibility play; it attracts those that are conscious of and want to support sustainability initiatives. Whether it's in attracting professionals to work for Capital One or multi-family tenants to live in a LEED certified building, it's great marketing material for that demographic.

So, for not giving value to the risk mitigating action, is the impetus potentially on the tenant? If tenants don't really see this as a problem and are willing to live in riskier flood zones, then is this even a problem worth solving?

Well, I still believe so.

The only reason people are willing to live in riskier flood zones is because their actual costs don't accurately reflect the real risks, and a lot of this comes back to the NFIP, as we have already seen. So, of course this is a problem worth solving. We need to act now and address this problem by incentivizing smarter development because we have already seen that given the current industry dynamics, a change like this won't naturally happen on its own.

Capital One is a great example of a corporation taking a strong stance on issues similar to this and implementing a corporate sustainability policy. This is certainly a step in the right direction, and I hope to see more companies take this lead.

Mr. Green also mentioned that the recent Harvey Storm had brought some of these risks to his colleagues' attention, and it has opened up opportunities to pursue further natural disaster risk mitigating initiatives. This is a great example, however, of the current situation we find ourselves in.

Only the large storms like Harvey draw peoples' attention to this issue and, for many people, it's often too late to do anything about it. For Capital One and Mr. Green, who didn't suffer severe damages, they are seeing the damages caused by Harvey and taking a proactive approach to begin to address risk mitigating initiatives now before it is too late.

Although the CDP and LEED don't deal directly with coastal risk mitigating initiatives, the wide spread use and popularity of these programs may help pave the way for our future. The program's ability to gain the support of institutional investors only helps to further their mission and shall be seen as a bench mark for which coastal risk mitigating initiatives will one day reach.

In the latter half of the book, we will look closely at exactly how the techniques used for gaining industry acceptance of the LEED program can and already are being used as a template to help jumpstart the real estate industry's awareness of impending coastal risks. It's clear currently that institutional investors don't stand behind coastal risk mitigating initiatives the same way they do with CDP or LEED. So, we will also look further at incentivizing and determining how to make these initiatives economically sustainable for everyone.

PART IV

MARKET CASE STUDIES

CHAPTER 6

HOUSTON WE HAVE A PROBLEM

"It's the poorest that are most vulnerable to climate change whether in this country or in developing countries." Paul Kirshen, Civil Engineer and Professor at the University of Massachusetts Boston School for the Environment

So far, our analysis of this problem has revolved around understanding the private real estate industry's development practices, the NFIP, LEED initiatives and the difficult task of achieving good, fast, and cheap results. We have established a core foundation of general knowledge around the issue of climate change and coastal development.

Now, however, our analysis will go beyond a generalized

scope to focus on the unique circumstances presented in specific at-risk markets. Through analyzing specific at-risk markets, we will gain a better understanding of the growing threat climate change poses and cover a broader scope of circumstances specific to each market. This will eventually allow us to develop a regiment or solution that can be widely applicable to different coastal markets.

The three markets will we analyze are Houston, New York City, and Boston. We will start in a market that has been hit recently. Houston has made headlines following the devastating 2017 hurricane season. Our analysis will then move to New York City, which has been identified as one of the most at risk coastal markets in the age of rising sea levels and climate change. We will finish our analysis 200 miles north up the coast to Boston, MA, where we will examine a market that is not as vulnerable as New York City but is already proactively addressing the incoming threat climate change poses to the coastal real estate market.

HOUSTON I: FEET ON THE GROUND

Upon graduation, Amanda Lee was not interested in pursuing a career path similar to her classmates.

Instead, she wanted to make an impact. Amanda Lee may not be your typical Hoya who heads off into a long career in politics, finance, or consulting. Instead, she wanted to have an immediate, material effect by serving those that need her expertise the most: the victims of natural disasters.

Before graduating from Georgetown, Amanda served as an emergency medical technician for Georgetown's Emergency Response Medical Services. In her final semester on the hilltop, Amanda interned for the Hazards & Climate Impacts Research center in Washington, DC.

It is through these personal initiatives that she gained unparalleled experience, preparing her for Mother Nature's wrath.

After graduating in May of 2017, Amanda went to work for FEMA as a Planning and Geographic Information Systems (GIS) specialist. Amanda spent her first few months as an analyst at the command center where she worked primarily with a program called Collector, a FEMA platform used to mark geographically damage and destruction in high risk areas.

The purpose of her work was to "identify geographically where the congregated damages occur so we can properly deploy the correct resources most efficiently to hard hit areas," Amanda notes.

She spent time taking calls from people in need, looking at all the incoming data, and developing the maps used to allocate resources effectively. However, to no one's surprise, as Harvey was making its way up the Gulf Coast towards Texas, Amanda knew her days in the command center were numbered.

Once the storm hit, Amanda was ready to help.

While the work continued to pile up back in the command center, Amanda was pulled from her position and sent into the field, where real people in real time desperately needed her.

After personally speaking with Amanda, I was extremely impressed with her background and experience prior to entering the field, but nothing could have prepared someone for what she was about to experience next.

On August 25th, Harvey made landfall near Corpus Christie, Texas, roughly 200 miles south of Houston. The eye of the hurricane passed right over Port Aransas and Rockport, Texas.

Immediately, Amanda was rushed to the front lines, positioned to work out of Port Aransas on Mustang Island, a small strip sticking out into the gulf from Corpus Christie.

Being sent to the eye of the hurricane, "I expected the worst, but it was beyond any expectations I had. The destruction there was the most severe in the area."

Fully submerged cars, boats pushed up onto land, and fallen trees provided a surreal and awakening experience for Amanda's first day in the field. In the first few days after Amanda arrived, the water began to recede and they could begin their efforts.

As a disaster survivor assistant, "I went door to door doing live needs assessments, finding those in critical condition and making sure they received the proper attention. I made sure people weren't living in unsafe dwellings that had been tagged as unstable, and, most importantly, I made sure that everyone I met registered with FEMA to ensure that survivor benefits were distributed to the right people."

Amanda recalls that a lot of people she met in the lower socio-economic neighborhoods had not registered for assistance because they did not properly understand what it was. Amanda had to continuously explain that it was disaster response aid put together for victims of the storm because people often mistook it for welfare.

Over the next few days, Texas fought for its life in the dark.

Harvey knocked down over 5,000 electrical distribution poles and 300 transmission structures, while damaging another 200, leaving over 220,000 people without power. Amanda remembers seeing American Electric Trucks (AEP) *everywhere*; "not only were there AEP trucks from Texas, I saw plenty of power trucks with Kentucky and even Ohio license plates," two of the other large regions serviced by AEP.

People were answering the call for help from across the country, and it paid off big for Texas. As of Sunday, September 3rd at 4pm, AEP reported that power had been restored to 178,000 of the 220,000 people affected by Harvey.

While power was beginning to be restored, Amanda's work was far from over.

Given that the eye of the storm hit this area, nothing really fared well. Destruction was everywhere. Walking down neighborhood streets, some of the houses had their entire front faces ripped away, exposing intimate family settings to the power of Harvey's wind and water. On the other hand, there were plenty of homes that appeared to be unaffected, but once you stepped inside, you could really see the devastation. "You couldn't walk or move anywhere because the ceilings had all caved in due to flooding on roofs."

On Mustang Island, no one could hide from Harvey.

Walking through neighborhoods, it was easy to identify the socio-economic classes by the recovery efforts already underway. There are some wealthy neighborhoods on Mustang

Island with large mansions that people often rent out on Airbnb or time-share as vacation homes. In these neighborhoods, it was clear that work was underway not only by the house owners, but also the housekeepers and contracted labor. Large piles of debris sat at the ends of driveways. There were plenty of local construction workers in the area, but the demand for work was simply too much. Further labor was contracted in from surrounding states similar to the electrical work to meet this demand.

When approaching these homes, Amanda spoke with a number of the local workers, and immediately signed them up for federal assistance. They, too, delayed applying for the FEMA aid due to confusion over the program. This delay meant that they had to wait eight to ten weeks because the four storms to hit the Gulf during the late summer months had crippled FEMA's response time.

Amanda recalls that back in the offices, FEMA was dealing with files of paper work coming all at once from Florida, Puerto Rico, Texas, and parts of the U.S. Virgin Islands. Through further conversations, Amanda learned that many of the workers doing debris pickup and construction were from Rockport.

While some of the destruction in Rockport was even worse than in Port Aransas, because the homes there were not built as well, most of the housekeepers and laborers started work in Port Aransas immediately. They mentioned that working right after the storm and not working on their own homes was the only option because they needed the money to repair their own homes.

These blue-collar workers "all neglected care of their own homes, where water was still sitting in some of the lower levels, because they couldn't afford to restore." They needed to return to work immediately because they were unable to afford any homeowner's insurance or flood insurance and were left waiting weeks for federal assistance.

This problem is not specific to the Port Aransas and Rockport areas, however, as the lack of flood insurance throughout the impacted area continues to be a hot topic across media outlets.

After spending nearly two months in the field, Amanda recalls, by the time she left, "it was clear there was construction going on at a lot of the houses in the upper-class neighborhoods, and homes were on the verge of being rebuilt. You could see, however, by looking at the amount of destruction still present in some neighborhoods both a month and two months after Harvey, what the economic prospect of a certain region was."

Only two months out from Harvey's landfall, the devastation is still very raw, and the focus remains on restoration.

While many restore, others decide to move on.

A number of families Amanda met with that owned vacation homes in Port Aransas planned to sell their properties. They simply decided it was not worth investing more into their property given the vulnerability of this area and their current knowledge about climate change. Should more people follow their lead and take Harvey as a sign of Mother Nature reclaiming parts of her ocean?

For those who can afford to, it certainly seems wise, but for those who can't, the cycle of restoring and waiting for the next storm may continue. In order to break this cycle for those who remain, once families and homes are stabilized, the focus must shift to mitigating future risk.

There is no denying that Houston was adversely affected by Harvey. Through my own research on natural disasters in the past and in observing the media's coverage of Harvey's impact, I realized, however, there is a tendency to pin these major storms to the largest cities in the region, often neglecting the smaller communities that faired far worse. It is for this reason I share Amanda's story with you.

HOUSTON II: EXACERBATING THE FLOOD

"I voluntarily decided to purchase flood insurance, and our

house didn't flood during Harvey. If it didn't flood during Harvey, at this point, I don't know if we'll ever flood hahaha." Christian McMurray

Christian McMurray is a long-time native of the DMV area. After practicing as a lawyer in D.C. for over a decade, he and his family decided to pack their bags and move to Houston in 2014.

In the weeks following Hurricane Harvey, I was able to catch up with Mr. McMurray to understand first-hand what life felt like during Harvey "inside the donut hole."

This is what Mr. McMurray refers to as living in the center of the three large beltways surrounding Houston, the State Highway 99, Beltway 8 and Interstate 610. To those that are familiar with the I-495 DC beltway, just for a size comparison, the Highway 99 beltway in Houston would encompass both Washington, D.C. and Baltimore, MD. Point being, Houston is huge!

Living in Virginia, Mr. McMurray had standard homeowner insurance with State Farm. After relocating down to Houston, however, State Farm refused to write a policy in Harris County.

"We had a relationship with them for 15 years in Virginia," Mr. McMurray details, "but when we moved over the summer of 2014, they said they were not writing policies in the area because it was hurricane season and the risk of loss was high."

This took Mr. McMurray by surprise but brought the concern to the front of his mind and promoted him to look further into the issue by studying FEMA's flood maps when looking at open houses. Through his research, he found that if you are in the 100-year flood plain, you may spend $4,500 a year in additional flood premiums, and, if you're outside that, you may pay $450 a year. It's really the 100-year flood plain outlined by FEMA that triggers a requirement for flood insurance.

Another surprise Mr. McMurray discovered when moving was the drastic difference in lot coverage ratios and lack of responsible urban planning in Houston. In Virginia, Mr. McMurray had a specific lot coverage ratio, which simply is a percentage of lot that is allowed to have a structure like a building. In his former residential zone, Mr. McMurray was only allowed to cover up to 45% of the lot with structure.

After moving to Houston, he realized the ratio is closer to 80%. This really begs the question, "do we really want to allow people to build these large homes on smaller lots where the runoff will now go into the streets as opposed to draining through the land on the lot?" That's the state of development in Houston today. Local legislature has limited the restrictions on development and zoning laws in the hopes to bring in more business, creating a real estate development free- for-all. That's the unique character of Houston.

As highlighted in the quotation above, fortunately Mr. McMurray's house was not flooded during Harvey and the remainder of his street fared well. Mr. McMurray attributes this to the investments his neighborhood made in new sewer runoff systems following tropical storm Allison, which brought floodwaters affecting over 45,000 homes. "Following that storm, they tore up the streets and made the drainage pipes a lot bigger."

These pipes now carry the water away from Mr. McMurray's neighborhood, down to the bayous, and eventually out to the gulf. This is a clear example of how added capital expenditure in mitigating future flood risks paid off in a big way.

Unfortunately, plenty of Mr. McMurray's friends in neighboring communities did not fare as well.

"A number of my colleagues at work were not in flood plains but were still impacted by flooding, especially with the release of the Addicks and Barker reservoirs."

On September 1st, 2017, officials began to release water from the two large reservoirs located in west Houston after they reached capacity. Officials did this in anticipation of more rain and to prevent uncontrolled releases at the spillways of the reservoirs. The Addicks reservoir continued to release 7,000 cubic feet per second (CFS) of water while the Barker released at 6,300 CFS. Those rates continued for around three days, affecting the surrounding neighborhoods where a number of Mr. McMurray's colleagues live.

In a news press release on September 1st, Jeff Linder said, "Some 3,000 homes near Addicks reservoir and 1,000 homes near Barker are inundated due to the water release." As the water continued to rage through neighborhood streets, eventually making its way to the Bayou, it caused the Bayou to reach an all-time high of 62.7 feet, surpassing a previous peak record of 61.2 feet.

In addition to Harvey, a significant number of Mr. McMurray's colleagues have been repeat victims. The Tax Day flood of 2016 and Memorial Day flood of 2015 each poured 15 and 12 inches of rain, respectively, over Houston in similar ten-hour windows. Because many of his colleagues were not within FEMA's flood plain, most of them did not have flood insurance.

After three years in a row of being flooded, not only may FEMA's flood maps need to be re-assessed, it's also certainly time for the local counties to invest in similar storm water management upgrades that occurred in Mr. McMurray's neighborhood. If this doesn't happen soon, at what point will these residents say enough is enough? After three years in a row, I think there is no coincidence, and many of the residents who have the means should consider relocating to higher ground.

From owning a policy and through helping the few colleagues who did hold flood insurance, Mr. McMurray learned that the policies usually account for $100,000 for possessions on the first floor and $250,000 on the dwelling itself. After being inside some of these destroyed houses where "you are ripping out all the drywall, insulation, and pulling out the electrical wiring that got wet, you basically have to rebuild an entire house," Mr. McMurray recalled. For a large ranch home in one of these neighborhoods ranging from 2,500-4,000 sq. ft. your insurance policy will only go so far.

When asked what the toughest part about the recovery efforts have been, Mr. McMurray noted the inability to find a decent contractor, driving construction costs up further. "To rebuild inside the beltway following the storm, given the high demand for contractors, you could easily spend over $200 a sq. ft."

Therefore, even if you had flood insurance, it would only help you rebuild a fraction of your home. So, owning the policy helps to mitigate costs, but it is simply not enough.

"It's not so generous to the point where it makes anyone not worry about the costs associated with flood damage," says Mr. McMurray.

This is because of how poorly run the entire Federal Flood Insurance program is. They provide very cheap flood rates that do not accurately reflect the risk associated with these properties and can therefore only give out small claim policies, leading to the current bankrupt state in which it stands.

So, what has all of this flooding really meant for the residential real estate market in Houston?

It would be easy to say there has been some type of collapse, but Mr. McMurray doesn't think this is the case.

"Individuals will be gutted on the value of their home, but, given the high demand for living within the beltway, the value of the lots will still remain."

Even after three years in a row of flooding, given the large amount of traffic and desire for people to be within the loop, individuals' innate desire to be centrally located will not change. For those that remain, it's imperative to build up to mitigate the risks associated with further low-land flooding.

One reason Mr. McMurray thinks the flood maps may need to be reassessed is because there is some belief in Houston that, to some extent, subsidence has taken place in certain areas of the city.

"As people drink from the natural ground water that certain areas have, the ground may have settled lower and led to increased flooding. I would say that right now, no one is going to build in the Meyerland area of Houston without building on an elevated platform."

Again, this only refers to those that can afford the means to do so. The real question lies with those that cannot afford insurance and are unable to move elsewhere.

HOUSTON III: UNEXPECTED FORTUNES

In my conversations with Mr. Roger Pratt, a senior advisor at the Elite International Investment Fund and an adjunct professor at Georgetown University, he was able to provide me the contacts I needed in order to include the perspective of an asset manager and third party operator to fully understand how these disasters affect the industry from all perspectives.

Mr. Pratt was able to put me in contact with Stephanie Nascimento, the Senior VP of operations for Alliance Residential Company. Alliance is one of the largest private U.S. multifamily companies, and, with Ms. Nascimento covering the South Central and Southeast portfolio, she was able to provide a first-hand experience of the impacts of the latest storms to hit the area.

Ms. Nascimento covers central Texas portfolio, consisting of over 22,000 units spread across the core markets of Houston, Dallas, San Antonio and Austin. The Southeast section extends all the way up to Nashville, TN, and the Florida markets consist of an additional 7,000 units under Ms. Nascimento's watch.

While Irma did cause some serious concern for Ms. Nascimento and her staff, the majority of damages they experienced came from Harvey in Houston.

One of the hardest things Ms. Nascimento had to do was ask people to leave properties.

"We have had physical assets encountered where we have had to de-occupy the building because it was unsafe; we have had to endure the costs of saying to our residents you need to vacate."

These costs for Alliance not only include the extra labor and construction needed to repair the assets but also the cost of lost rent payments for the entire building. While scale and scope of damages differed for many real estate companies, for Alliance's Houston portfolio, in particular, "there were either assets with very little damage, or ones that were extremely impacted; there was nothing in between," said Ms. Nascimento.

There were two assets, in particular, that Ms. Nascimento had to vacate. While both of the properties were severely damaged by flooding, neither of them was within a designated flood zone and, therefore, did not carry proper flood insurance.

This situation is not unique to Alliance as "many of the apartment dwellers, developers, and single-family home owners were not in classified flood zones and did not have the proper insurance," Ms. Nascimento noted.

This, again, brings our attention to questioning the validity of FEMA's historic flood maps. Both of the properties vacated, Ms. Nascimento mentioned, were flooded due to the release of the Houston reservoirs. So, certainly the maps need to be re-drawn, but the question becomes: shouldn't they also include those neighborhoods that are vulnerable not only to initial flooding but also to overflows and releasing of the reservoirs?

In terms of their tenants' personal belongings, Ms. Nascimento said it's "hit or miss" whether they bought renters insurance to cover their goods. "We require our tenants to have renter's insurance for general liability over the entire community, but it's up to them if they want anything for their personal belongings, and no one really secures flood insurance as an apartment dweller, especially if you're not in a flood zone," said Ms. Nascimento.

Due to the severity of the storm and damages, like the two Alliance assets, there has been a spike in evictions and high delinquency because people were having trouble paying their rent. Ms. Nascimento pointed out that this caused the courts to be more tenant friendly than usual and caused some stresses for the asset's owners.

In addition to vacating unsafe buildings, Ms. Nascimento noted that there was a significant amount of time where the fallout of tenants was felt and people were struggling to catch up from the storm and couldn't afford to keep living in the property. When asked if they attempted to relocate any of the vacated tenants to undamaged properties also under their ownership, Ms. Nascimento said of course. But, as a third-party manager with multiple different ownerships across different assets, the challenge was dealing with different rent structures between properties.

"There is some price variation across the properties we manage, and so we had to have some sensitive conversations with tenants. In some instances, the owners waived the application fee or deposit amount to come up with a supportive approach for new tenants, but we could not guarantee vacated tenants were going to get the exact same rent at their new location."

When I asked how long she expected it would take to get the damaged properties back up and operational, Ms. Nascimento assured me that each would take considerable time. "We don't expect to get our club house and amenities up and running until May at one of the properties," she said. They will attempt to release starting in February only the second and third floor units of both projects because the first floors are being completely re-done and they anticipate those to take nine to twelve months to be ready.

Now after having fallen victim to the Harvey floods, I wanted to understand to what extent the risks of flooding are considered in the company's due diligence process and whether they anticipate these damages to change anything moving forward. Ms. Nascimento assured me that, as a company, they take their due diligence process seriously, they avoid building in high risk flooding areas and they do not have any assets in high risk flood zones currently. This was a very unique situation with the release of the reservoir and given that the assets are not in the flood zone; Ms. Nascimento anticipates that her team will not do anything drastically different in the future.

My major concern is that, in their due diligence process, they are assessing risk based on faulty flood maps that don't accurately depict the risk of flooding for their properties so clearly demonstrated by the release of the reservoirs. If the real estate managers don't change their behavior and continue to rely on FEMA's flood maps, the impetus then relies on the government and FEMA to reform its maps and supply accurate flood zone predictions that take into consideration the impact of global warming.

When asked about mitigating risk beyond what they pay for insurance, Ms. Nascimento drew a comparison to the LEED initiatives. "For our portfolio, I don't see any added benefit from mitigating risks further," she said, because over the years, they have not seen a significant return on our LEED investments.

"I haven't really seen in the market that we are getting an extra $X in rent because we say a property is LEED certified and has met all the qualifications. In some of our deals, we haven't even seen a premium on sale because of the LEED certification."

While investing in mitigating the future risk of flooding may pay off if there is another large storm, Ms. Nascimento is more concerned with the public's current perception of the property. Mitigating future risks may ease some tenants' concerns, but, overall, it could draw more unwanted attention that the property did flood. Who would want to live somewhere knowing that it just flooded? Ultimately, the question becomes whether the investment in mitigating risk is enough to offset the public concerns over a property that just flooded.

As has been described, Alliance has certainly suffered damages from the impact of Hurricane Harvey. What might surprise you, however, is that overall the Texas portfolio benefited from Harvey's storm.

After initially discovering this, I was at a crossroads. I listened to the various ways in which Alliance benefited, which I will get into next, but I was left thinking to myself on the other end of the line, "I am trying to write a book on how devastating natural disasters can be to coastal real estate communities and discover ways to catalyze change in our approach to developing in these areas. How is this possible if these storms are benefiting the private industry? Why would they change anything in the future if they have net gains from these disasters?"

At the time of my interview with Ms. Nascimento, I felt these were all valid questions I was considering and consequently, I began to worry about the direction of this book. After reflecting further on my entire research process and my conversation with an economist from the National Association of Realtors, it is obvious that some communities suffer during these events, causing others to benefit from that suffering. This tradeoff is ultimately helping to solve the problem of increasing risk of loss by causing people to migrate from highly risk prone areas to lower risk prone areas. So, while Alliance experienced losses with its most risk prone assets in the path of the reservoir's release, it overall benefited from its less risky assets that were not affected by the reservoir's release.

Apart from the two assets they had to de-occupy, Alliance highly benefited from the fallout of other communities and single-family homes that were impacted by the storm because a lot of people needed a quick place to live. With the way the storm hit Houston, the damages were pocket driven.

"For the people impacted inside the beltway, they really want to stay in the same general vicinity where they currently are because that's where their schools are, their employment, and their life, so they will begin to move to the un-impacted nearby communities."

The Alliance properties in neighboring communities with vacancy immediately saw their occupancy shoot up to 100%. "I had one property that was previously offering some healthy concessions, one month to six weeks of free rent, and we were sitting at 88% occupancy before the storm. After the storm, we jumped to 100% occupancy and released without any concessions given the high demand." This jump in occupancy caused the properties rental income to increase over 30% within a short time period following the storm.

At first, I expected most of these new leases to be short-term leases because people simply needed a little time to get back up on their feet. Ms. Nascimento noted that she, too, expected short leases. But, since the anticipated period of recovery given the amount of labor and construction needed for these homes is expected to take seven to ten months or longer to be livable again, their property was able to get full-term leases out of this disaster.

From an operator's perspective, Ms. Nascimento knows that running her properties at 100% occupancy is not going to last long-term, so attempting to find where the happy medium will be once the entire market has settled and appropriately budget that for the next and coming years will be her biggest challenge.

HOUSTON IV: NEGLECTING CODES

"99.9% of all buildings out there are existing buildings. You're talking about a lot of master planning infrastructure and buildings that were completed prior to 2005-2010. We are living with decisions made really in the mid-to-late-1900's, and, when you talk about overcoming 50 years of developments, it's going to take a long time to have properties redeveloped and incentivize new standards that minimize the impact on the infrastructure and allow them to absorb some of the climatic changes. It could take another 30-40 years for that building stock to overturn, and that, of course, is economically driven." Chris Garwood, DCS Design.

After attending the D.C. Metro Area's Premiere Green Building Symposium, the 3rd Annual D.C. Sustainability Summit presented by DCS Design, I was extremely fortunate to follow up and connect with Chris Garwood, Vice President at DCS. The day-long conference celebrated the region's commitment to green design and focused on increasing awareness about sustainability in the commercial real estate community. DCS design is a full-service master planning architecture and interior architecture firm with a special emphasis on sustainable design practices.

My conversation with Mr. Garwood began with his telling me the quotation mentioned above.

Mr. Garwood was one of the first individuals I was able to interview for this project and immediately after hearing that quotation, Mr. Garwood paused. I immediately began thinking to myself, "Jeez. How am I going to be able to provide anything tangible to cope with a problem of this magnitude?"

I was born in 1996, so a lot of the decisions and issues we are now facing as an industry and society were decided before I was even around. A lot of the design problems go back 40- 50 years, specifically with storm water management infrastructure noted by Mr. Garwood. His familiarity and experience within the DC metro area has made him a local expert on sustainability design.

"Some of the facilities built in the area of Tysons, VA are at their maximum capacity, and what this means is that if the storm water run-off regulations had not been changed, we would have had to build an entirely new additional storm water runoff plant to deal with the overflow."

Considering the amount of tax dollars and infrastructure dollars that go into a project like that incentivized locals to search for an alternative option. So, what ended up happening in Fairfax County and Tysons specifically is that they mandated that future construction must be able to retain the first inch of runoff water for a storm in order to ensure the existing storm water infrastructure does not get overwhelmed. Creating this mandate allows the water from each site to be released over a longer period of time.

"This has been a great example of how local jurisdiction works to reduce the need for expanded infrastructure with storms in the short term."

Mr. Garwood and his team at DCS Design recently worked on a project in Arlington County, where they designed a storm water retention bowl. The system was only built to hold a couple inches of rain at any one time; anything more than that and the system would be overwhelmed.

"To put in place a system that could handle upwards of five to six inches of rain at one time would take up a large garage level or two. It reaches a point where it makes the project impractical."

This is not only because of the space a system of that size would require on site, but also the costs associated with putting in place large infrastructure of that nature. The development community is really not taking on initiatives to handle large storms of this nature because of the constraints on space and budgets.

The impetus has really been on local jurisdictions to push for changes in storm water management, and even then, "it's focusing more on what's happening 360 out of 365 days and not on the couple of days every year or couple of years where there is a significant natural event."

Given the large percentage of properties built prior to the one-inch storm water management initiative that was put in place, D.C. is still playing catch up, and will continue to do so, in terms of providing the adequate infrastructure needed to deal with large storms and excess rain.

This problem, however, is not unique to D.C. With its local jurisdictions mandating LEED certification for properties, which in spirit addresses the need to reduce the amount of hardscape in urban design and investing in proper storm water runoff infrastructure and an abundance of green roofs,

D.C. is thought to be at the forefront of this issue. There are larger cities and coastal communities that are in far worse condition, including Houston.

When asked to give his expert opinion on the condition in Houston, Mr. Garwood quickly responded, "It was all a nightmare waiting to happen."

During the master planning in Houston, "they basically ignored the LEED spirit and paved over everything. So, if you have impervious surfaces everywhere, there is nowhere for the water to go."

That being said, it's important to provide further context of the environment in Houston. Overall, the building footprint in Houston is likely less than half the surface area of the city itself, Mr. Garwood speculates. In that case, there is very little real estate developers could do to prevent the catastrophic flooding that occurred with regards to their own individual storm water management systems. Instead, most of what needs to be done is in the city streets and sidewalks, and "you have to be willing to use more pervious concrete and asphalt."

While these materials are currently available, the reason they are not being used is because of the premium to be paid for them. You have to dig down deeper and set a deeper gravel bed than traditional materials in order to produce a more pervious surface.

However, even if a similar rainstorm happened in D.C., with all our LEED certified properties and green roofs, you would still have catastrophic results. When it was all said and done, in the four days the storm sat still above Houston, from Saturday, August 26th to Tuesday the 29th, one trillion gallons of water fell. That amount of rain is equivalent to how much Houston typically receives over an entire year.

In Mont Belvieu, a small city a couple of miles east of Houston, an unbelievable 51.88 inches fell over that same time period. "Any time you have as much rain as you did in Houston," Mr. Garwood concluded, "there is no infrastructure that is going to be able to handle that."

This should not, however, discourage us from improving our infrastructure to withstand greater levels of flooding. The damages seen in Houston could have been significantly reduced if there was not a complete disregard for the LEED spirit and proper building code in the city's master planning and if more infrastructure was in place to deal with storm water runoff. Like it was mentioned before, we are playing catch up. Mother Nature is showing no signs of slowing down, so we must continue to innovate and incentivize proper sustainable development.

HOUSTON V: THE EVOLUTION OF SUSTAINABILITY—LEED

Mr. Garwood recalls that in the early days of LEED, there was a push to used recycled materials and reduce energy costs that was driven by local jurisdictions; "but that being said, they kind of jumped out in front of where the market place was."

The new LEED certifications mandates by certain jurisdictions required construction to be done in more sustainable ways, but the materials to build different weren't as readily available. Developers were then left paying a premium to use these new materials. Once materials and the manufacturing industry grew around these new regulations, they were able to provide more competitive products, allowing sustainable construction to become more affordable.

What I find most fascinating and exciting about LEED is they continue to progress and push the bar higher. Mr. Garwood supports this point by saying, "What first happened is the jurisdictions start pushing LEED, then what follows is that the people who write the building codes start to write code to match some of the goals with LEED." This means that what you needed to accomplish in order to get LEED certified five years ago is now what you have to do to be code compliant; therefore, it is not considered an extra expense but, instead, a mandated expense to build sustainably. This basically means that we are in a cycle where LEED comes out with a new version and they raise the bar; within three years, the code usually catches up, and so LEED sets the bar higher.

The adoption of LEED and writing of code to comply with its goals by different jurisdictions has allowed this stepping stone process to occur. It has helped jumpstart the sustainability initiative in an industry that more or less neglected sustainable building practices for the past century. Arlington County is a great example of taking this initiative. Mr. Garwood is very proud of this adoption by local jurisdictions and notes, "there has not been a development in the past ten years that has not been at least LEED silver certified. It has become a regular practice." It's inspiring to see the LEED certification process be a catalyst for change in the real estate industry and to consider the different aspects of this initiative that can be applied to catalyze, in a similar way, a shift in the industry's approach to climate change.

CHAPTER 7

NEW YORK IS WORTH SAVING

NEW YORK I: BEFORE IT ALL

Earlier this summer, I was in New York for a leadership symposium. As part of the weekend event, a group of 20 of us were all taken on a river cruise for dinner called the Spirit of Boston. The cruise departed from Chelsea Piers and made its way down the Hudson, looping past Battery Park and the Statue of liberty before making its way back north. Being in the middle of the river and looking back at the city provided me a very different perspective contrasted with the view from my hotel room. On the top deck, my eyes gazed out over the railing following the water until they met land. Or, the more appropriate terminology would be until they met the concrete piers because they are likely nowhere close to where the original coastline of Manhattan Island once stood. It was incredible to see from this perspective how crowded and overbuilt the island appeared.

The creation of Lower Manhattan is something that likely should have never happened. How the city grew from a swampy area to the busiest and most developed city in the country is an astounding story and needs to be reviewed when looking at the risks associated with flooding and storm surges in the area. In 2009, Eric Sanderson delivered a TED Talk detailing just that.

Over the past 15 years, Mr. Sanderson has worked alongside his colleagues on what they call the Mannahatta Project. The project's focus is to discover what Henry Hudson would have seen on the afternoon of September 12th, 1609, when he first sailed into the New York Harbor. Mr. Sanderson opened his talk with the following quote:

"As the moon rose higher, the houses melted away until I noticed the island once flowered for Dutch sailor's eyes." F. Scott Fitzgerald

During the historical development of the city, it is important to note that NYC was the first mega-city, defined as a city of ten million people or more, in 1950. But what did the city look like before all the world-renowned skyscrapers and distinctive skyline? Mr. Sanderson provided his guests the painting below.

"For those of you who are from New York, this is 125th street under the West Side Highway. (Laughter) It was once a beach."

Mr. Sanderson was then able to take a number of different maps and georeference them by placing them on top of a modern-day grid map of the city.

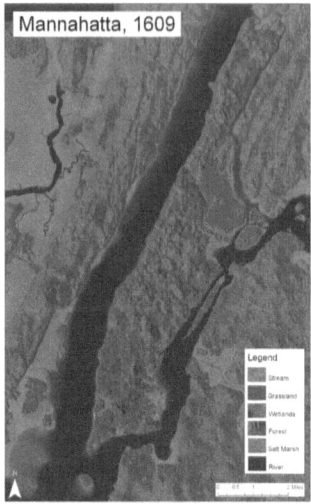 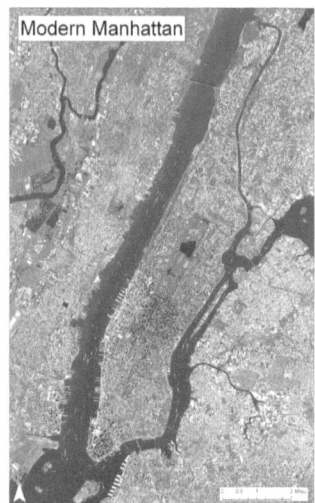

Through this exercise, Mr. Sanderson was able to pinpoint what once stood in the block-by-block geography that people are familiar with today, where people now live, work, and play.

It turns out that before the city became what it is today, it once had 55 different ecological communities. Mr. Sanderson notes that, "On a per-area basis, Manhattan had more ecological communities per acre than Yosemite, than Yellowstone, than Amboseli. It was really an extraordinary landscape that was capable of supporting an extraordinary biodiversity." The area was once filled with streams, meadows, forests and wetlands that are now all paved over. Now, all we are left with is the rolling topography of Central Park as it comes up against the abrupt and towering Midtown Manhattan skyscrapers.

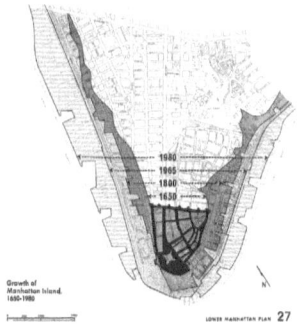

Before the infill apparent from the maps provided, the Hudson itself used to be a lot wider. I recently watched a documentary about the infrastructure nightmares during the 9/11 recovery efforts when they feared that underground walls keeping out the Hudson River were damaged and may cause underground flooding. We attempted to build where buildings were never supposed to be. As is apparent from the images above, we filled in parts of the Hudson River for development that now put billions of dollars of real estate in danger.

NEW YORK II: ORGANIZING CHANGE

On Tuesday, November 14th, 2017 I awoke to the reverberating sound of sirens as they passed by my New York City hotel room and continued off into the distance. The sun was piercing through the ends of the heavy drapery as it hung over the window wall of my Eurostarz hotel room. I had checked in the night before and immediately went to bed. Now, I mustered up the strength to creep over to the drapery and peek into the light.

All I saw was blue.

Blue skies and blue water.

I watched as the glimmering waves of the East River splashed up against the city's edge, only feet from where the base of this building stands. This was not the first time I visited New York City, but every other hotel I have stayed in had been surrounded by other large buildings and the view out of those windows did not go farther than ten feet until my eyes met another building shooting up into the sky.

Staying at the Eurostarz on the city's edge in Lower Manhattan was different.

In that moment of standing next to the glass wall of my room, I felt more exposed than ever. To my right, I could see where the East River meets the mighty Hudson in-between Battery Park and Governors Island and flushes down into the Atlantic. I thought to myself how vulnerable and exposed I was if the water ever happened to flush the other way.

Ironically, I was visiting the city to compete in Cornell's ninth annual International Real Estate Case Competition. Instead of focusing on one of the largest real estate markets in the world and addressing its vulnerable state to incoming natural disasters, we focused on a retail property in downtown Madrid.

The next morning, after finishing the competition and returning to D.C., I was able to connect over the phone with Malcolm Bowman, an oceanographer and trained physicist engineer who studies storm surges in the New York area; he was on a train headed towards NYC.

That morning, Mr. Bowman invited me to a meeting that he would be attending that included oceanographers, meteorologists, engineers, marine biologists, architects, city planners, and the Port Authority of NY and NJ; unfortunately, I had already returned to DC.

At the beginning of our conversation, he emailed me the agenda for the day's meeting, which included a number of presentations from Bill Golden, head of the NICHI, as well as updates on a number of studies and projects currently underway in the NY-NJ metro area. I was fortunate to pick Mr. Bowman's brain over the next 30 minutes about the progress they were achieving.

It was clear from the start that the politics associated with a large initiative like this were very complicated. At this point, Mr. Bowman and his colleagues have been working with the Army Corps of Engineers, looking at various ways for possibly protecting the region. It's standard to perform a cost-benefit analysis and the engineers quickly lose interest if the costs far exceed the benefits.

However, Mr. Bowman mentioned that the ways in which they compute benefits is rather simple. "They look at infrastructure that has been damaged or destroyed. Suppose you have a warehouse on the waterfront that has been flooded. They ask, what are the costs to repair it and what is the benefit of repairing it? They don't look at situations in which we build a barrier system to protect the city for another 100 years that provides a lot of new opportunity for development along the coastline which was previously too risky to develop earlier."

While both Mr. Bowman and I agreed that to calculate benefits in this manner could be a lot trickier and complicated, it would provide a better understanding of the true long-term cost-benefit relationship of large infrastructure projects.

When I asked Mr. Bowman why they do not take this approach, he responded, "It's all governed by regulation. The Army Corps cannot do the more sophisticated analysis of the loss/gain opportunity because of the regulations they abide by which limit their ability to do an extensive analysis."

While I took this to mean it simply reflected a lack of adequate funding to produce a more sophisticated analysis, what really left me scratching my head was when Mr. Bowman said, "They even have concern over their appearance at the meeting. They don't want to show favoritism to one group or company."

At this point, I could only think it had something to do with the current administration's stance on climate change, given that the Corps is a federal agency.

Despite these constraints, Mr. Bowman and his colleagues are attempting to work with them to take a more modern approach. Unfortunately, "the political faction is five layers deep and everyone has their own ideas of what to do. The politics create an extremely inefficient process and catalyst for change."

In order to make progress and achieve tangible results, it's imperative that governors and mayors in this area work together because Mother Nature knows no politics. After speaking with private developers and understanding further their investment strategies in coastal communities, I understand how critical a role your own time horizon plays into determining how much can be accomplished.

When asked how this dynamic might translate into the political environment Mr. Bowman has observed, he said, "The political system is short term by nature, and politicians are always thinking about the next election cycle." In the context of mitigating future climate change risks, this means that for a governor or mayor that will only be in office for the next couple of years, there are few incentives for them to take initiative or cooperatively work together on a solution if they feel it's unlikely something catastrophic will occur while they are in office.

On the other hand, Mr. Bowman noted, "at universities, you have the privilege of being able to think long term, 100 years ahead." It is this discrepancy between time horizons of organizations that severely limits the ability to achieve progress. Personally, I think the city of New York is on borrowed time, and, if something is not done about it now, it will be too late by the time the next storm hits.

At this point in our conversation, it was clear both Mr. Bowman and I were becoming frustrated talking about the political friction that has plagued progress. Either in an attempt to lighten the mood or provide some glimpse of hope for me to include in my reflection, Mr. Bowman said, "There is one instance in which people have been able to move past the politics and achieve success, and that's in New Orleans. Since Katrina, the Army Corps worked to build a system around the city, and they did it rather quickly. If the political will and funding is there, these things can happen."

I immediately thought to myself, "is that really what it's going to take to achieve progress? Is a Katrina-type storm the only thing that will catalyze a change in our behavior?" In the words of Celeste Hammond, a professor of law and director of the Center for Real Estate Law at the John Marshall Law School, "I keep wondering how many disasters do we have to see before we think about it. It's really a serious problem."

Mr. Bowman continued by emphasizing how they are not going to sit around and wait when he said, "we are all working very hard to try and make sure that is not the case. We are fundraising right now and have a proposal for $3.1 million ready to go for further feasibility studies. But between the economics, the environmentalists, the fisheries, The Army Corps of Engineers restrictions on construction, and the politics, it's a huge challenge."

NEW YORK III: BUILD THAT WALL!

Before Mr. Bowman got off the train to head to his meeting, I asked him one final question: "I understand there are many different types of barriers around the world protecting coastal cities, but is there any one barrier in particular that relates to the unique circumstances here in New York and that could be used as a model for further proposals?"

"Probably the most similar barrier to what is required in New York is in St. Petersburg, Russia." Mr. Bowman responded,

"They started building a beltway like you have around Washington, D.C. It's an elevated six-lane highway that is built on the Neva River delta. Construction began in the 70's, but, after the union collapsed, it sat uncompleted for years."

Halcrow, an international engineering company headquartered in Lower Manhattan, was selected to complete the remaining construction of the Neva River Barrier in St. Petersburg, Russia. Once completed, the barrier stretched over 20 kilometers long, creating a protective wall around St. Petersburg filled with gates to allow efficient ship traffic and drainage flow out of the Neva Bay.

Only a year after the wall's completion in December of 2011, it got its first real test as heavy wind and rain from a winter storm pushed a wall of water into the Neva Bay. The wall stood strong, protecting the lives and real estate within St. Petersburg.

"That's when we learned," says Graeme Forsyth, a civil engineer and a director of Halcrow in Scotland, "that this barrier really works."

After identifying a number of similarities between the situation in Neva Bay and New York Harbor such as the water depths, sea bottom and adjacent spits of land, and potential surge heights, Halcrow proposed a similar project for New York called the Outer Harbor Gateway.

The one large difference between the two projects, however, is that the proposed one in New York, at eight kilometers long, is only a third of the length of the one already built in St. Petersburg. This creates a unique opportunity to "plug and play" with the parts of Neva's barrier needed for New York, says Forsyth.

"You could take St. Petersburg apart and plug in the bits you want to create a new configuration for New York."

The proposed Outer Harbor Gateway runs from Sandy Hook, NJ to the Rockaway Peninsula in New York. It accounts for three large gates that would allow ship traffic as well as twenty-two smaller "sluice" gates to further allow proper flow and drainage of water in and out of the bay. All of the gates would be able to close when a storm surge approaches. According to early Halcrow estimates, the cost for such a barrier would run around $6.5 Billion.

One of the major concerns about this potential project is that it would create a bathtub environment within the New York Harbor. If the gates were closed when large rainfall and runoff occurred during a storm, the Harbor would begin to fill up like a tub and would flood coastal communities from within the wall.

However, Jonathan Goldstick, an engineer and Vice President

at Halcrow, addressed this concern by saying the Neva Bay, as well as New York Bay, are "large enough of reservoirs to take heavy rain and river flow," without allowing something like this bathtub scenario to take place.

"It's simply the best role model for NYC," said Bowman, "and we have a debate among my colleagues about a potential barrier and the multiple purposes it could hold, like a light rail between the airports of Kennedy and Newark or an interstate toll road."

Like Mr. Bowman mentioned, the Outer Harbor Gateway could include a six-lane highway resting on top of the barrier similar to the one in St. Petersburg. This potential highway would connect central New Jersey to the Queens borough of New York and to the rest of Long Island, which could reduce traffic congestion passing through the heart of the city.

While I myself am not a frequent commuter within the city, when mentioning this proposal to a number of my friends from the city, they believed a road coming from NJ to LI and bypassing downtown would have a material impact on millions of NY and NJ commuters, significantly reducing the congestion downtown.

Not only would a proposal like this make sense for commuters, it could also aid in what will likely be the most difficult aspect of the project, which is acquiring funding. If a highway project was approved, part of the costs could potentially be provided by federal highway funds or through private investors if it served as a toll road.

If executed properly, Goldstick is confident the project's engineering studies, design and construction could all be completed within a decade, providing a reliable protective barrier and safe New York Harbor for generations to come. Unfortunately, though, Goldstick feels this is unlikely to occur because he's "never seen the federal government or local regulators move quickly enough to allow this kind of timing to happen."

Furthermore, there is fear that the ecological and environmental studies required for a project of this magnitude could slow the project down immensely. To properly determine how such a barrier would impact life within the harbor could take as much as ten years on its own, Goldstick suggests.

* * *

During my conversation with Mr. Rutter from Tishman Speyer, who was introduced back in Chapter Three, he concluded by suggesting a number of risk mitigating activities that could be done by NYC developers in the future to help reduce the vulnerable state of the market to future climate change. Potentially raising the lobby and moving important utilities and building infrastructure to a raised podium off of the ground floor would help reduce the risk of damages in the case of a flood.

Mr. Rutter also acknowledged the possibility of a large barrier wall to protect the entire city. By putting in a wall, you would have to believe that, at some level, by decreasing the risk of something going wrong you would be increasing the value of the asset. Therefore, if you put up a wall and property values go up, then the government could potentially get private funding for the infrastructure from the companies that own large properties impacted by the wall. "To what extent would the value be increased," is the real question Mr. Rutter believes we should be focusing on. If insurance gets reassessed and the assessed risk is decreased after the wall is built, then your insurance costs would decrease and people would attribute value to that. When I asked him how much return would be needed to make this a feasible solution to the problem of funding, he replied, "It's such a theoretical question. It just doesn't really work that way. You can't just say 'I need the value of my property to increase by x% in order for me to fund this project.'" Public private funding is a very tricky and difficult task to accomplish. Unless it's something along the lines of increasing the BID tax for Lower Manhattan, getting a significant portion of the real estate community on board to fund a public private project would be extremely difficult.

NEW YORK IV: LESSON LEARNED FROM SANDY

While Mr. Bowman has addressed the need for a large, mega structure to mitigate future risks and further damage associated with storm surges, it is clear there are a number of things that have continued to slow that process down. Similar to the suggestion Mr. Rutter has about developers to take action by raising podiums, there have been a number of advancements made following super storm Sandy to mitigate future risks that should be acknowledged.

Curtis Wahle, a Commercial Construction/RE Development Executive, who has spent the majority of his professional life building in Manhattan, informed me of a number of significant changes in the Building Code that have come as a reaction to the damages from Sandy.

"New York City reacted faster than any municipality has to the devastating damages from Sandy. Within months after the storm, they were able to implement significant improvements to the Building Code in an attempt to mitigate future damages," Mr. Wahle notes.

The International Building Code (IBC) is a Code developed by the International Code Council (ICC) that has been adopted as a base Code in most jurisdictions within the country. Most cities adhere to the IBC, a State Code, or some iteration of the two, but New York City is one of the only cities with its' own building Code. This allows the City to specifically address the inherent and unique risks within the market and particularly in Lower Manhattan.

One regulation set forth in the NYC Building Code mandates that the City adhere to a higher datum than what is set forth by the Army Corps of Engineers. A vertical datum is a base measurement point from which elevations are determined. Historically, the National Geodetic Vertical Datum of 1929 was the building standard. Now, the North American Vertical Datum of 1988 is the national standard. However, each of the five boroughs in NYC now adhere to their own vertical datums set forth in the Building Code.

"Raising the datum above the Army Corp of Engineer's datum means that we assume sea level rise is higher than what the Army Corps tells us it is, and that all new construction must adhere to this higher datum" says Mr. Wahle. "All buildings constructed within the flood zones in the five boroughs are required to have a flood barrier system to the height of the NYC Code's Datum High Water Mark (100-year flood)." This means the building must essentially be water tight up to this level to mitigate the effects of possible flooding.

"In our designs, we must include flood panels and doors that can be installed in front of all openings, and window systems below this level must all be sealed systems. Below ground garages must have full flood doors and panels at their ramps, etc.," notes Mr. Wahle.

Another extremely important and little-known fact to most outsiders of the City is that NYC has a combined sewer system. This means that there is one system that carries both storm and waste water.

"The system is ancient and undersized, especially in Lower Manhattan." Mr. Wahle noted that during Sandy, "the system completely failed because it was unable to handle the deluge of rain water and not to mention the surge given its limited capacity– there was simply nowhere for the water to go."

As a result of this failed system, which is very similar to the problem experienced in Houston, NYC builders are now required to build storm water retention systems on each property site. This typically consists of huge tanks on the sub levels of the development. These tanks are designed to hold the storm water on property and gradually release it into the combined sewer system to regulate the amount of inflow into the limited capacity system.

"Considering the building boom since Sandy, this has the potential to have a significant impact during the next superstorm, and the potential impact will only increase as time goes on and more new construction is completed," Mr.

Wahle concludes.

As Mr. Wahle has noted, there has been significant progress made following the devastations of superstorm Sandy. This progress is helping to paint a brighter picture for the future of New York City, and specifically Lower Manhattan. These changes to the Code will hopefully prevent the kind of property damages the City sustained following Sandy.

NEW YORK V: GLOBALLY COMPETITIVE COASTLINES

"People are saying, 'let it go back to the beavers,' but New York City is worth saving." Bill Golden, National Institute for Coastal and Harbor Infrastructure

Soon after deciding to take on this topic in my attempt to write a book, the first number of articles I read that addressed the risks of coastal flooding due to climate change and how we can work to mitigate them included insight from Bill Golden. His name kept appearing on article after article about the subject, so I knew right away he was someone I needed to interview.

A couple of weeks later, I was able to connect with Mr. Golden and, from our conversation, I was able to understand how important he is in addressing the large problem of our inadequate coastal infrastructure in this country.

Long before his sustainable efforts took form, Mr. Golden found himself applying to Georgetown Law School. After being admitted and postponing his acceptance for a year, he did not end up attending at all because he "wanted out of Washington." While Mr. Golden never had the chance to play with Jack the Bulldog on Healy's front lawn and will never truly know the meaning of "Hoya Saxa," I did not let this get in the way of the important business at hand.

During his time in Washington, D.C., Mr. Golden was recruited to work for the White House and played a role in the creation of the EPA.

Now, Mr. Golden is founding president of the National Institute for Coastal and Harbor Infrastructure (NICHI). According to its website, the NICHI is a nonprofit educational organization that is committed to building a national coalition of private and public interests to advocate for a "national interstate coastal infrastructure system that integrates, enhances and funds local, regional and state coastal climate adaption plans."

The NICHI is responsible for organizing the first national conference of community leaders from areas affected by sea level rise. The group has met several times at this point, but the likelihood of a unified response remains low.

In response to this, Mr. Golden begs the question, "But what do you do with a city that represents 400 years of people's blood, sweat and dreams, and the financial and arts capital of the world? People are saying, 'let it go back to the beavers,' but New York is worth saving."

With little positive progress coming from some of these meetings, Mr. Golden has decided to take an unorthodox approach in addressing this issue. Mr. Golden has invested in two of the world's sixteen Nantucket lightships that remain: Lightship 612 and Lightship 613, the last two ever made.

While Mr. Golden and his wife are able to spend summers in my home state of Rhode Island on board Lightship 612 in Newport, during the offseason, the boat serves a more educational purpose while in the New York harbor. Mr. Golden began our interview by telling me about a trip he had just taken onboard his refitted Nantucket Lightship 612.

"We just finished up a tour around the port of New York with elected officials, government committee leaders and the media," he said. The purpose of taking tours like these is to help foster an accurate understanding of the vulnerable state of vital coastal infrastructure as well as commercial and residential structure.

It helps to emphasize this point from the perspective of being on the boat, where you can clearly see that "even on clear days, the water laps at the edge of Lower Manhattan just a foot below the sidewalk."

Back before Lower Manhattan became the roaring global financial center of the world, people began to build along the waters because of the human attraction to aesthetically pleasing and recreationally abundant coastal communities.

Hundreds of years later and that attraction has only grown.

Mr. Golden points to the increasing risk associated with these prized communities by referring to the National Academy of Science report that predicts 500-year storms are now expected to happen every 25 years. This is in light of the fact that 12 of the 13 worst storms in the past 100 years have occurred since 2004.

The problem is further exacerbated by the fact that many of the buildings built decades ago were designed with 100 year lives but did not anticipate the rise in sea level and increase in storm surge we are experiencing today.

A significant number of our coastal buildings are in serious jeopardy. This has become a national economic and security issue, not just from physical damages but also from the impact a regional disaster can have on an economy.

This problem stems beyond just coastal real estate.

In an attempt to bring further awareness to the issue, Mr. Golden and the NICHI are initiating a coastal educational tour in the spring of 2018 onboard Lightship 612.

The trip will include touring 48,000 miles of US coastline to collect stories of storm surges and to document the effects of rising sea levels, focusing specifically on the 7% of U.S. landmass that is treated by rising sea levels. Golden is hopeful this outreach and documentary work will bring more attention to the fact "that another Sandy storm could cost just New York alone $90 billion" if the proper mitigation efforts are not taken.

If a seawall were in place and working, Mr. Golden says a hurricane, "even with a potential storm surge of more than 25 feet, it would feel like just another windy day in New York City."

When I asked Mr. Golden how he believes we got into this mess in the first place, he pointed to our history of inadequate responses to rising coastal threats by saying, "the federal agency could significantly impact the implementation of public policy goals, but we have an antiquated agency that focuses in terms of skills and the traditional concepts of mission that don't reflect the new realities."

What Mr. Golden really means by this is that the federal

government has been stuck in a focused cycle of disaster relief and repair. If the focus of your policy is to respond to disasters by funding repair to get communities up and running as soon as possible, then you are going to continue to build the same way with little reduction in the vulnerability of new structures.

What is needed now more than ever is a shift in policy from disaster relief and repair to planning and investing to protect our aging infrastructure against rising sea levels and increased storms.

Mr. Golden's vision is for an integrated national coastal infrastructure system that includes the entire coastline, addresses the new realities and adapts on a local basis. In order to gain awareness and traction, Mr. Golden's strategy has not focused on just defending coasts and avoiding damages, but instead has emphasized the opportunity for expanded development. "We have the opportunity to create a situation in which we can spur on a whole new age of economic and community development along the coasts."

In order to achieve an outcome of this magnitude, the right amount of funding necessary to spur this coastal renaissance must be present. At this point in our interview, Mr. Golden again looked to the history of this country, and, specifically, our past major infrastructure investments. Significant times of economic development and growth accompany times in which we have invested in our great nation's infrastructure, including in the 1800's, when we invested in canals and railroads, and the 20th century, when the federal highway system was advocated by Dwight Eisenhower. Mr. Golden noted that, in order to properly assess the current state of coastal infrastructure, we will need a federal investment of this magnitude.

While this will require significant costs upfront, Mr. Golden believes it will ultimately pay off big time in the long run. Preventing further the effects of climate change on coastal infrastructure will not only protect coastal communities, but it will simultaneously create new opportunities to invest in commerce, transportation, energy generation and communication along the coast. Mr. Golden brought up a very valid point when he said, "sooner or later we are going to have to do an investment of this nature to protect ourselves anyway, so we might as well use it as a multi-beneficial investment that creates a wave of economic development."

Mr. Golden is confident in the lasting benefits created by an investment in coastal infrastructure and warns that a failure to do so may place our country at a global disadvantage in the growing age of a global economy.

Currently, only two to three ports in the U.S. can accept the large Pan-American class vessels, and the new, even greater vessels coming through the Suez Canal will not be able to fit in any of our coastal ports. "We likely aren't going to build a significant number of new federal highway lanes," says Mr. Golden, "so we need marine transportation to take tens of thousands of containers off the roads and deliver them by water." In a global economy, being able to accommodate the ships of the future will become increasingly important. We need to invest now to protect our future.

Apart from investing in the proper marine infrastructure to support a first-class global economic player, when asked what his solution was for coastal real estate threat by the increasing effects of climate change, Mr. Golden offered two answers. The first is to retreat and run for the hills. He noted that there are plenty of strong advocates who believe certain places should be bonded and returned to the ocean or wetlands. His second solution is a stay and fight approach.

He believes a combination of both provides the best approach. We need to look at everything and decide "ok, this is where we retreat, this is where we take a stand. Then, once we decide where to take a stand, we work through a plan to maintain and stabilize the economic viability of the communities."

Mr. Golden firmly believes that some places, which potentially includes Miami, must be abandoned because there are simply no technically effective means of protecting them. It's obvious that certain areas should not be built in now, and it may be necessary to acknowledge the limited life of structures that were built and intended for much longer lives in coastal communities. Mr. Golden concluded by saying, "There are beaches that should never have been built on, and many of the structures will have to eventually be given up, bought out or destroyed when insurance is no longer available."

CHAPTER 8

BOSTON PUSHES FORWARD

"Something has to be done. Whether it's this or more small-scale solutions, Boston has to come up with a strategy to protect itself against increased coastal flooding." Paul Kirshen, Civil Engineer and Professor at the University of Massachusetts Boston School for the Environment.

BOSTON I: MODEL TOWNS

The worst hurricane to hit New England was back in 1938; it caused nearly $4.7B in damages. It is safe to say that, given the coastal real estate development since then and in accounting for inflation, if a storm of that magnitude struck today we would be looking at catastrophic damage costs. In a response to that storm, three coastal communities, but not Boston, took it upon themselves to protect against future flooding by building storm surge barriers in the 1960's. Before we dive deeper into the logistics of a wall in Boston, let's first look at those projects that preceded it.

STAMFORD, CT

The first of these projects is the Stamford Hurricane Protection Barrier:

Looking back on historical flooding, losses for the small community of Stamford, Connecticut following the storm of 1938 totaled around $6 million. Nearly 15 years later, the region was struck again by Hurricane Carol in 1954, resulting in $3.4 million in damages, of which, the U.S. Army Corps of Engineers (The Corps) estimated that $2.9 million could have been prevented if an effective barrier was in place. This was enough to prompt Stamford in 1965 to protect the nearly 600 acres of commercial and residential communities vulnerable to storm surges. The project took just under four years and cost around $14.5 million. As of 2001, The Corps estimates that more than $38.4 million in coastal flooding damage has been prevented by this barrier.

Specifically, the Stamford Hurricane Protection Barrier consists of three separate parts, all working together to protect and stabilize the Stamford community. The first part is a 2,8500-foot long earth fill dike with a 90-foot wide opening at the East Branch of Stamford Harbor to allow traffic in and out of the harbor. The second part is a 1,3500-footlong concrete wall and a 2,950-foot earth fill dike that protects the west branch of the harbor. The third part is a 4,400-foot earth fill dike that protects the Westcott Cove in Stamford. All three barriers have pumping stations to adequately deal with different water levels and proper drainage of sewage and debris from within the harbor.

Stamford, CT

PROVIDENCE, RI

Providence is also home to one of the three storm barriers in

New England. Being from Rhode Island, I am extremely familiar with this barrier. As a middle schooler, every morning at 7am I would ride the school bus into Providence, passing over the east/west bound I-195 bridge known as the I-way, which runs parallel to the Fox Point Barrier only inches downstream. I gazed out through my waking eyes at the large gates of the barrier and never once did I consider how valuable that structure was to the stability and longevity of the downtown district.

Now, nearly ten years later, I am reflecting on these experiences as a young boy and praising my hometown for taking initiative and for leading the way for future coastal infrastructure projects. Construction on The Fox Point Protection Barrier began in 1961 when the city said enough is enough after water depths of up to eight feet were recorded in the downtown commercial district. The hurricanes of 1938 and 1954 caused $16.3 million and $25.1 million in damages to the area respectively.

Similar to the Stamford project, this barrier took around four years to build and cost $15 million. The barrier consists of a 700-foot long, 25-foot high concrete structure spanning the mouth of the Providence River where it meets Narragansett Bay. The concrete structure has three large gates that are left open to allow small vessel traffic into the downtown river, but it can be closed in the event of a coastal storm. A 780-foot eastern earth fill dike and a 1,400-foot western earth fill dike flank each side of the concrete structure, providing further protection for the city and its surrounding neighborhoods. These earth fill dikes were built on pre-existing roads and therefore have three vehicular gates that can close when a storm surge is expected.

Providence, RI

NEW BEDFORD, MA

The third New England barrier is found 50 miles south of the Boston Seaport area in New Bedford, Massachusetts. The vulnerable area represents 80% of all the land affected by the 1938 and 1954 floods. Similar to the other two projects, the New Bedford Hurricane Protection Barrier started construction in 1962 in response to the 1938 and 1954 hurricanes. It took four years to complete and cost around $18.6 million. The barrier now protects 1,400 acres of surrounding communities, consisting of heavily developed industrial and commercial corridors.

The barrier consists of three parts. The first part is the main 4,500-foot long earth fill dike extending across the New Bedford and Fairhaven harbor that stands 20-feet tall and has a 150-foot wide gate to facilitate commercial and residential vessels. The main barrier extends on land, stretching 4,600 feet along Rodney French Boulevard East to further protect the low-lying harbor bank. The second component is the 5,800-foot long earth fill Clarks Cove Dike in New Bedford, extending around the north and east side of the cove. Last is a 3,100-foot long earth fill dike that runs easterly along Alton Street in Fairhaven. The Corps estimates that as of 2011, the New Bedford Hurricane Protection Barrier has prevented $24.1 million in flood related damages.

New Bedford, MA

For Boston, the time is now. Boston is one of a number of

cities that requires serious attention with regards to future flooding vulnerabilities. Fortunately, there are a number of programs and projects in place that have already initiated the discussion around this issue. The most notable of these programs is Climate Ready Boston, which proudly states on the front page of its website, we are "an initiative to develop resilient solutions to prepare our city for climate change." Boston is determined to adequately prepare and defend itself in the new age of climate change. They don't want to join a growing list of cities that have experienced serious damages because of ineffective preparations.

BOSTON II: MODEL PROJECTS

Each of these New England communities offer insight and guidance into the difficult task ahead for Boston officials to protect their beloved city. Each of these projects cost less than $20 million and, while there is not a cost estimate yet, experts agree a potential Boston Harbor barrier will likely cost tens of billions of dollars.

When asked about the potential price tag associated with a project in Boston, Paul Kirshen, a professor of climate change at the University of Massachusetts Boston School for the Environment and National Climate Assessment author, responded, "This project won't go ahead unless the economic benefits far exceed the cost of the project." However, with $80 billion in real estate assets and 90,000 residents to protect in Boston, the costs associated with building a wall can still be justified.

This is certainly possible if the proposed barrier can provide further stability and opportunity for economic expansion within the Boston Seaport area. So, what would a wall mean for the future of real estate development in Boston?

Bill Golden, who was introduced in Chapter Seven from the NICHI, insisted, "If we make strategic investments, not only do we avoid the downside of death and devastation, we also provide opportunity for communities with economic development." Therefore, when assessing the costs associated with a project of this magnitude, it is important to weigh the opportunity costs associated with not funding a barrier project. If Boston remains a sitting duck to further storm surges and rising floods, this could reflect poorly on the long-term profitability of the region. Each year that passes without a barrier in Boston, the flood risks for real estate developments gradually increase.

David Levy, a professor of management at UMass Boston said in a Bisnow interview, "We're asking ourselves, what are the broad impacts on business, particularly if the city gets a reputation for being vulnerable." Failure to take action and actively mitigate future risks associated with flooding waters will not only increase the risk and insurance premiums for real estate in the harbor area, it will deter future development and adversely impact the real estate market.

There are only a few infrastructure projects in Boston's history that rival the potential barrier in size and potential impact. The first that comes to mind is the Big Dig, which re-routed the I-93 downtown corridor underground and provided new tunnel connectors to Logan airport. This project created more efficient traffic flow through the city and made room for what is now the Rose Fitzgerald Kennedy Greenway, a park spanning the length of the old downtown corridor.

Personally, I have driven through the Big Dig plenty of times and have experienced the easy flow straight through the city, whether it's catching a flight at Logan airport commuting to my high school in New Hampshire from my home in Rhode Island. During the summer months when I was home, I would often find myself taking day trips to Boston, walking from South Station along the Greenway, passing the iconic Boston Harbor Hotel on my right, and making my way up to Quincy Market for a bite to eat. With the lush park and manicured street scape, I have experienced first-hand the positive impact the Big Dig has had on the real estate assets abutting the Greenway.

Many still fear, though, the budgeting disasters associated with the Big Dig, which ended up costing the city $24.3 billion after it was initially estimated at only $2.8 billion. However, both Kirshen and I are confident this project's execution will be different and "the mandate for this project would be clearer than the Big Dig in terms of pure economics," noted Kirshen.

It is likely that the final cost of the Big Dig will be close to what a potential barrier surrounding the entire Boston Harbor might cost. Mr. Golden highlighted the magnitude of this project and its relation to the Big Dig when he said, "We need funding on the level of the interstate highway system." However, because of the looming concern around the Big Dig's mishandled budget, Mr. Golden drew our attention to his own Boston Harbor clean-up project as a better comparable.

Prior to Golden's efforts, the Boston Harbor was known as one of the dirtiest harbors in the nation due to the inefficient and dated sewage treatment infrastructure. The polluted waters created an unclean and uninviting environment around the harbor that was starting to hurt the region's economics. The value of real estate assets along the harbor would be in jeopardy if the issue was not properly addressed. During the same time as the Big Dig, Golden led the efforts for a $3.8 billion project that would update the Harbor's sewage treatment facilities in hopes of reducing the polluted water and creating a safe and clean environment for future economic and real estate growth.

Mr. Golden related the difficulties they've faced in gaining momentum and support for the barrier project to his experience with the treatment facility. "When we started out, we were told this is going nowhere, but we did it. People came from everywhere and it happened. That's the value in speaking truths to power." It is his perseverance, demonstrated through this project and his experience with the NICHI, that left me confident in his ability to move forward with the vision for a protective barrier.

Now, the Deer Island Sewage Treatment Plant protects the Boston Harbor against sewage pollution flowing from Boston's sewer systems. As you can see in the image below, the plant has significantly reduced the amount of pollution in the Boston Harbor and has therefore had a tremendous impact on real estate values, providing a safe and clean environment for billions of dollars in real estate development following the project's completion. Today, this project is recognized as one of the greatest environmental achievements in the nation and should be used as a benchmark for the future storm surge barrier.

Similar to the sewage pollution situation, the longevity of real estate along the Boston Harbor is once again in question due to rising sea levels. Boston now has the opportunity to build a revolutionary wall, protecting billions of dollars of real estate and providing a safe environment for further real estate investment and economic expansion in the future.

While the goal of such a project is to reduce the impact of high tides and stop future storm surges, in doing so, it would reduce the overall flow of the water in and out of the harbor and bring into question the ecological impacts a barrier may have. That is why it is important to have a significant number of openings and gates, similar to the three barriers identified earlier, as well as numerous pumping stations to ensure proper flow and drainage of inner harbor waters. With billions of dollars put into cleaning the Harbor, it's imperative the proper research and design is applied to this barrier project to ensure the water remain clean while creating a protected harbor from storm surges. With Golden's personal interest in maintaining a clean harbor, I am sure this aspect of the project will receive the appropriate attention.

BOSTON III: THE FUTURE OF THE SEAPORT

Paul Kirshen is helping to catalyze a shift in the city's approach to climate change. Currently, his work is focused on Boston's funding initiatives through different foundations and the National Oceanographic and Atmospheric Administration to look at developing security based adaption strategies for metropolitan Boston.

To deal with the threats currently facing Boston and the Seaport area, Mr. Kirshen outlines three major strategies:

1. Protect for the future-Build the wall
2. Accommodate for the future-Let the flooding occur and flood-proof properties
3. Retreat-Move away

While the third of these strategies is the only solution that will 100% reduce the risk of loss and damage associated with coastal storms and flooding, for many reasons, this is highly theoretical and will be very difficult to achieve in reality. For this reason, Kirshen suggests that an appropriate response to the current threats will include a combination of the first two strategies. In order to be done effectively, this will require everyone to work together to come up with district wide risk mitigating solutions.

In addressing the first strategy, we have already compared the potential price tag of a wall to other historic Boston projects, but what exactly would a wall look like in Boston?

An article published by the Boston Globe in 2017 titled, *As seas rise, city mulls a massive sea barrier across Boston Harbor*, outlined a few different proposals.

The first wall proposed would be a small barrier stretching from Castle Island to Logan Airport. This was the smallest of the proposed barriers, which means it could be built the fastest and cost the least. However, this barrier would only protect the inner harbor and would still leave much of South

Boston vulnerable to coastal flooding.

The second barrier would stretch from Deer Island, across Long Island, to Moon Island in Quincy. This would be a slightly larger project, and would protect more of South Boston, but would still leave areas south of Quincy vulnerable.

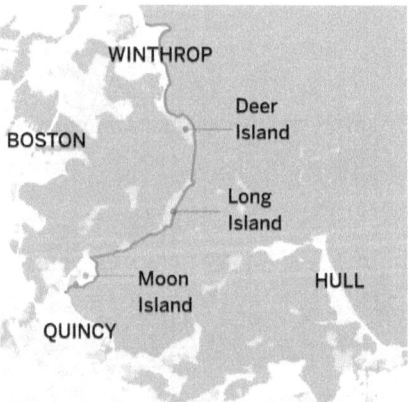

The third barrier is one proposed by Bob Daylor, a private

engineer in Boston. He coined his proposal the Sapphire Necklace, and it would involve a series of dikes stretching from Deer Island to Hull's Telegraph Hill.

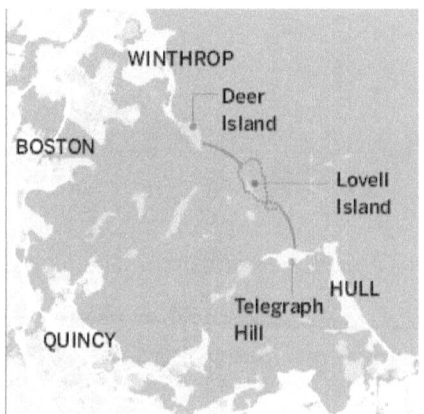

Similar to Daylor's proposal, the fourth and largest proposed barrier would stretch the entire four miles between Deer Island and Hull's Telegraph Hill. At low tide, this proposed barrier would rise 20 feet above the water level. All of the barriers proposed would contain openings to allow water flow and boating traffic to pass through that would then close when a large storm surge is approaching.

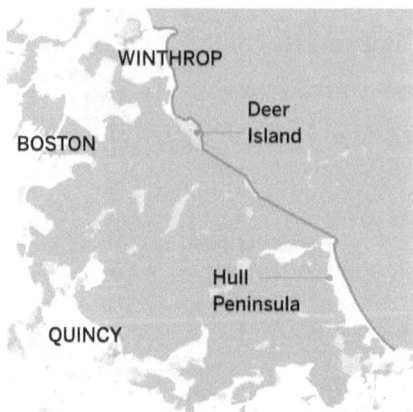

As the Boston Globe article notes, the magnitude and potential cost of this final barrier can be compared to similar projects in New Orleans, Venice, and Rotterdam.

As has already been mentioned in Chapter Seven, following Hurricane Katrina the Federal Government spent over $1 billion over five years to build a protective barrier in Lake Borgne. In Venice, Italy, where coastal flooding continuously threatens the city, they spent $5 billion to construct a 1,200-foot barrier. This barrier was first proposed in 1970 but has taken decades to complete. In the Netherlands, nearly a quarter of their land is below sea level. After a large flooding storm hit Rotterdam in the mid 1900's, the city spent close to a billion dollars in 1990 to extend a preexisting 66-foot barrier another 600 feet. The new barrier now stretches across the Nieuwe Waterweg waterway.

In an interview with the Boston Globe, Martien Beek, a deputy program manager at the Dutch ministry of Infrastructure and the Environment, said, "We haven't had any flooding since then. This is a solution that has proven itself, and it could work for Boston as well. All major coastal cities that have big tidal movements should be considering this."

Boston can look to these other projects around the world as successful models when building their own protective barrier. However, the wall proposed in Boston will likely end up being significantly more expensive and potentially take decades to build.

This is where the cooperative, district wide approach becomes necessary. In addition to the long-term work being done by local government officials, experts such as Mr. Kirshen and engineers say, "real estate developers should be taking their own adaptive responses in the meantime because it will take plenty of time for a wall to be approved and built."

Within the Seaport area, a significant portion of Boston's commercial real estate assets are like sitting ducks, awaiting the next big storm. Most of the development is relatively new; "there are a lot of high-tech buildings in the area, but most were still built without the consideration of climate change in mind," said Kirshen. One way in which developers can reduce the damage associated with coastal flooding is to focus on the second strategy Mr. Kirshen outlined in order to accommodate for the future and flood-proof properties.

This can be achieved by designing properties where the key utilities are built higher up, allowing the first floor to flood in the event of a storm. Or, the sites landscape can be designed to include a berm around the property, both of which will make the building itself resilient. While these measures may be required for some of the properties most vulnerable to coastal flooding, Mr. Kirshen emphasized that, given all the tunnels and mass transit in the Seaport, "working together to protect it all at once should be the long-term focus. It'll be cheaper than everyone working individually."

BOSTON IV: ADDRESSING THE THREATS

Currently, there are a number of initiatives in place to increase awareness about the threat of climate change for future developments along the coast. Developers are still building large structures in Boston. In order to build, you need to receive a permit from the Boston Planning and Development Agency (BPDA) and fill out a climate change questionnaire for them. This questionnaire asks developers what they are going to do about potentially higher temperatures, more rainfall, and more flooding in the future. While the developers respond to this questionnaire, Mr. Kirshen highlighted, "at this point, it's just an informational tool for BDPA. They can encourage developers to be more resilient but they can't force them to."

While this is a step in the right direction, a material change in developer's approach to climate change will likely only occur when it becomes mandated. The city is currently trying to change the building codes so they can force developers to build more resiliently. "That's a very difficult process," said Kirshen, "because at the moment in Massachusetts, the building code is dictated by the state code, and it's difficult for localities to set their own building code." Both Kirshen and I concluded from this that it is in the region's best interest for the city to come up with its own building codes to address the vulnerable state of real estate assets along the coast.

Mr. Kirshen then pointed me to the exceptional work being done with the Climate Ready Boston report, which has produced more forward-looking maps compared to FEMA's that effectively address the risk of rising sea levels and increased flooding.

"There are now plenty of flood maps out there that show what the flooding will be in the future, as well as plenty of consultants in the area well equipped on the topic, so there is no excuse for the developers not knowing."

As of now, the zoning has not been adjusted with these maps in mind, but conversations about doing just that are underway. In a year or so, Mr. Kirshen anticipates the city will have made some progress on this matter.

According to Mr. Kirshen, most of Boston's waterfront is now only about a foot above high tide marks. This is because most of Boston is located within filled in, low lying tidelands, which is apparent in the image below (all of the lighter color such as the airport indicates infill locations). In the coming decades, if no flood mitigating actions are taken, large sections of the city could experience regular flooding at high tide.

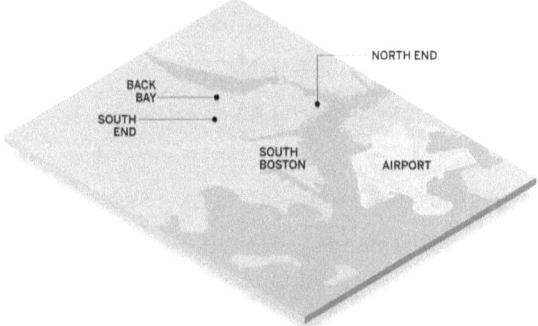

The main threats facing the Seaport area, although related in a number of ways, can be separated into two specific threats.

1. Extreme rainfall and the risk that it doesn't drain out into the harbor fast enough.
2. A bad coastal storm that brings a large storm surge into the Seaport area.

The major threat of extreme rainfall is due to the under designed drainage networks currently in place. "The systems were built decades ago and rainfall has increased since then, so the system is not designed to handle this new amount and there will be local flooding in the streets," said Kirshen.

If sea levels go up, which they are expected to do, this will exacerbate the drainage problem further because water flows from high to low elevation. The rate at which water flows out is proportionate to the height difference between water in the streets and water in the harbor, so if sea levels go up, water will drain more slowly from the streets to the harbor, making areas more prone to flooding.

The other problem is a storm surge in the ocean, which would cause significant damage if a protective barrier were not in place. A storm that brought both of these would be one heck of a mess, "but that's the reality of the situation. We must start preparing for the future now," said Kirshen.

When asked in further detail what solutions to these problems will look like, Mr. Kirshen responded with emphasis on creating a comprehensive, multilayered approach that addresses both green infrastructure and a wall.

Green infrastructure, with best management practices and low impact development in mind, will help to reduce the rate of runoff of rain in urban areas. The buildings themselves could have storm water management systems or "blue roofs" that are designed with the explicit intention of storing rainwater. This will only go so far in addressing the risks of a storm surge, however, and Mr. Kirshen said, "we can deal with the nuisance of rain flooding, but we must prioritize our approach to storm surges."

In addressing the need for a wall, Mr. Kirshen believes the toughest part is dealing with the uncertainty of climate change. "We don't really know if we'll have three feet or ten feet of sea level rise be the end of the century."

Given this uncertainty, it's important that our strategies are developed to be flexible. This is not a problem we can solve for just today's risks. We must be pro-active in our approach and develop infrastructure that can be adjusted overtime to the changing risks.

For example, "if we want to build a protective wall for Boston, we might build it five feet above sea level now but with a base that can potentially support 15-20 feet in a couple of decades, creating flexibility for the future." Creating a structure with flexibility will certainly take more design and construction time, so it's imperative that progress on this initiative takes place today.

BOSTON V: FUNDING

One of the major obstacles to furthering progress on these projects is funding. Following their respective storms, New

York, New Orleans, and Houston have all received federal funding to restore and rebuild. Boston is being proactive in their response to climate change, attempting to mitigate the risks before the next storm hits. Therefore, the same federal funding is not being provided, and they are trying to figure it all out on their own. "If we can't figure it out on our own, then we're going to have huge economic losses in the future," said Kirshen.

During the construction of a number of the New England barriers mentioned at the beginning of this chapter, a group effort was needed to acquire the proper funding. According to Mr. Golden, "local elected officials and local business associations got together and got the barrier funded. In the case of New Bedford, the city contributed, the state contributed and businesses contributed."

Furthermore, there is a lot of literature now on financing adaptation that goes beyond the potential of toll roads talked about in Chapter Seven. Mr. Kirshen is currently working on a project focused on catastrophic bonds, green bonds, and incriminate tax financing. The report will be release in February 2018, but one of his ideas is to set up a resiliency district in an area like the Seaport. The city will then borrow money to build a wall to protect that area based on the anticipated increase in value of the properties protected by the wall.

Therefore, the increased value of properties will raise the tax revenue generated in these areas and allow the city to pay off the wall. Mr. Kirshen said, "in the bond market, people borrow against increased tax revenue. There are schemes out there to finance this stuff," noting that this is a common practice that can be adapted to this specific situation.

Mr. Kirshen ended our interview on a high note by emphasizing a number of initiatives the city continues to work on, giving me hope for the future of this coastal real estate hub. I recommend all of you explore on your own time the Climate Ready Boston website, which should be used as a gold standard for all coastal communities vulnerable to the future risks of climate change. In the near future, a new report will be released detailing the potential costs associated with protecting the Seaport area.

PART V

THE KEYS TO LASTING CHANGE

CHAPTER 9

THE BENEFITS OF PRIVATE INSURANCE

Now that we have looked at a number of markets and analyzed the unique situations present in each, let's return to our discussion on flood insurance. Through my conversation with Mr. Doe of the NFIP subcontractor, I was able to better understand the NFIP in its current form and identify some of its key inefficiencies. While a number of projects suggested through our market studies will take significant planning and time to come to fruition given the economic state of the NFIP, reform is needed immediately and will help to progress risk mitigating initiatives in all U.S. markets vulnerable to coastal natural disasters.

If privatized insurance is the direction we are headed in, then what does that mean for policy holders?

Significantly higher prices.

This will likely cause a lot of commotion for current policy holders, but the increase is certainly justified and needed. The new policies will then reflect the actual risk associated with coastal properties instead of the subsidized government policies that currently provide a false sense of risk.

If private insurers became involved, they would charge very sophisticated rates. They would take into consideration things like recent construction, what neighbors have done to mitigate risk compared to you and what the storm projections are for the next 15-20 years, to name a few. All of the data collected would determine the price of the policy, which will be dramatically more expensive than what anyone is currently paying in flood prone areas.

To better understand what a privatized approach might look like, I reached out to one company leading the way in this field, Munich Reinsurance.

PRIVATE ROLE MODELS

Not everybody believes that climate change exists or even leads to the increase in natural catastrophes.

"Even if you don't believe it," says Pina Albo, past president of the re-insurance division of Munich Reinsurance America Inc. (MR), "they happen, so how do you prepare yourself for the future ones better than you did in the past?" Currently, Ms. Albo serves as a member of the board of executive management, and I was fortunate to connect with her to learn more about MR's efforts in this space.

At MR, they choose to recognize the implications of climate change and are very focused on the topic of resiliency and how to develop smarter because "we are not going to get rid of natural catastrophes," said Ms. Albo. While MR is one of the largest reinsurance companies, they don't necessarily re-insure all of the properties that are covered by the NFIP. If they are re-insuring clients that insure those properties, then typically "that portion of our coverage will be carved out and will be at such a level that just makes financial sense," said Ms. Albo.

Through my conversation with Ms. Albo, and in order to provide me with a more detailed understanding, she put me in contact with Peter Hoeppe, who is the head of the Geo Risk Research and Corporate Climate Center at MR. Once I connected with Mr. Hoeppe over the phone, he quickly explained to me that natural disasters are one of their core businesses; "for 40 years, we have had natural disaster scientists doing risk analysis in terms of damages," he said.

In addition to this, MR has done significant lobbying for where and how people should be allowed to develop in the wake of climate change. Ms. Albo said through their lobbying experience in London, they have learned it is difficult to enact a change in behavior. "If a political figure is up for re-election," she says, "they have a hard time telling their constituents how and where to build. There is a political component to the whole thing that is hindering progress." Political setbacks like these, however, have not derailed MR's efforts.

Within the industry, MR believes the privatization of flood insurance is a step in the right direction.

Ms. Albo stated, "if it were in the hands of the private industry, we would not see rates charged to policy holders that are so cheap and does not encourage them to fortify their homes or to not build at all." In its current form, the program doesn't encourage anyone to act smarter than before, causing a significant amount of damages to come from repeat victims.

Ms. Albo noted that, at the time of our conversation, there was debate in the industry surrounding this issue of whether repeat victims should still receive a subsidized policy. Mr. Hoeppe explained how important these conversations are to change our approach of addressing climate change when he said, "in the private world, if your house burnt down twice, chances of you getting insurance the third time at cheap rates are very low, and, therefore, it encourages certain behavior." It is imperative for us to look at how current policies are affecting and influencing certain behaviors and question whether these current behaviors are sustainable.

It's because of the passive behaviors currently being incentivized by the NFIP that make it an unsustainable program and have caused it to go bankrupt.

Continuing to fund this program in its current form is not sustainable for society, as well, because it does not encourage smart behavior such as not building right along the coastline or building smarter and stronger when you do.

If there were no federal sponsored programs, people would be forced to act differently because they would not have access to cheap insurance.

This, then, would accurately reflect the gamble of building along the coastline because one would know that, if it blows down, they are not going to be federally bailed out. Once this process becomes no longer federally subsidized, or as Mr. Hoeppe said, "Once the light switches on and people wake up," then people will be incentivized to improve their property. Maybe there will be some type of government sponsored program for people on the coastline that can receive tax deductions if they make improvements to their homes safety. Programs like this will effectively incentivize resiliency efforts that will ultimately reduce the damages associated with natural disasters in coastal communities.

HOW TO PROPERLY UNDERWRITE RISK

Once the industry is privatized, Mr. Hoeppe noted, "Insurers and reinsurers will look at risk mitigating factors before deciding whether to write a policy or not, and, if so, at what terms and price to charge for the insurance coverage." Although this type of practice is not currently a part of the NFIP, it is already incorporated into the work at MR.

As mentioned earlier, with over 40 years of natural scientists running risk analysis on natural disasters, MR has taken a special interest in this specific space. Mr. Hoeppe outlined the process in which they address these risks and noted that the first thing they begin to analyze is how frequent a certain event with a certain intensity is. For each submarket, they have models that predict how often a hurricane makes landfall at certain levels of intensity. The next thing they do is collect data on how vulnerable the structures are in the market, factoring in their past loss experience as well as changes in the building codes. Mr. Hoeppe reflected that in Florida, following Hurricane Andrew in 1992, the building codes were changed to a higher standard, which made a visible difference when Irma made landfall in Florida.

In the areas where the new building codes were enacted, "the losses were much lower after Irma than in the area where building codes remained unchanged," said Mr. Hoeppe. The third and final factor MR takes into consideration is the exposure, which means how many houses of a particular value are concentrated in a certain location. If a hurricane makes landfall, "we need to know how much of our hurricane portfolio may be affected," said Mr. Hoeppe.

Taking all three of these aspects into consideration helps make up MR's loss model, which will then tell them their expected loss and is used to calculate a risk output premium needed to cover the associated risk. This is a basic, simplified explanation of risk insurance and reinsurance.

One very important aspect which MR considers is adjusting for normalized losses, which takes into consideration the increase in development and population along coastal communities. This simply means adjusting the historical damages associated with natural disasters for the increase in value of properties and population along the coast to make losses comparable. Their methodology for adjusting is based on gap growth, which means that, as people get wealthier, the values exposed to hurricanes increases more than inflation; so, this must be taken into consideration in risk assessment.

Even after taking normalized losses into consideration, Mr.

Hoeppe identified that 2017 will set a new record level for normalized losses; "2005 with Katrina, Wilma, and Rita had been the record year until now, so the normalized losses have trended upwards, as well." This means that the increase in damages cannot be simply explained by social, economic, or demographic data. There is a high probability that the weather has changed and this means that climate change should play a role in future development within coastal communities.

GOING ABOVE AND BEYOND

In addition to their lobbying work, MR has also actively invested as a main partner in the Institute for Business and Home Safety (IBHS).

Located in South Carolina, the IBHS is a research institute that has one of the largest wind tunnels in which they can put an entire house and apply hurricane level winds up to category 3 to test roofing and other hurricane proof materials. The findings from this organization have also helped to improve building standards and avoid increases in damages.

One project, in particular, that MR has worked on in conjunction with the IBHS is a development along the New Jersey coastline that focuses on building sustainable and resilient homes after Sandy. I was unable to get a hold of Mr. Hedi, who is in charge of the project in New Jersey, but Ms. Albo explained that the project consisted of investing in new, risk mitigating construction best practices and materials. The project's purpose is to be an example of smart development in vulnerable areas and "to show people that it doesn't cost that much more to properly mitigate risk," said Ms. Albo.

Through this project, MR has invested in a brighter future for coastal developments in a time of increasing risk. During my conversation with Mr. Hoeppe, he directed my attention to an interesting study he had seen about the damages following superstorm Sandy. Manhattan had the most significant losses caused by Sandy in 2012 after the storm surge flooded the subway system. Mr. Hoeppe noted that Lloyds, the London based insurer, "has calculated that 30% of the Manhattan losses are attributed to sea level rise, and this is clearly an effect of global warming."

With the insurer's demonstrated interest in sustainable development, MR, in conjunction with IBHS, has created a free app, *Fortified Homes,* that is used by homebuilders to get tips and suggestions for mitigating risk on their property and to see how they can make their own house less vulnerable. It is platforms like this that will help contractors navigate the new era of smart development along coastal communities.

THE TRANSITION LOGISTICS

Now, we can all understand the advantages associated with a privatized flood insurance industry. However, one of the main reasons the transition to private insurers has not already happened is because the logistics associated with such a transition can be complex and difficult to address.

To learn more about the economic impacts climate change poses for the real estate industry and the economics associated with transitioning the flood insurance industry, I reached out to Mr. Yun, the Chief Economist with the National Association of Realtors (NAR).

After starting with the NAR in 2000 as a junior economist while still taking graduate classes, Mr. Yun worked his way up and was promoted to his current position in 2008. In his current role, Mr. Yun has spent the past decade overseeing a staff of 12 economic, statistics, and computer science experts to study the real estate market and determine how global changes may impact the industry.

Mr. Yun notes that while the real estate industry does not look out long term, the damages associated with climate change must be analyzed over the long run in order to prepare properly. Their work in this space consists of covering areas related to natural disasters and flood insurance, assessing the damage from natural disasters and determining how long recoveries of certain markets may be. Some of their work includes sending information to policy makers about the need of proper flood insurance, recognizing the fact that homes or properties that get repeatedly flooded should be priced appropriately to account for those actuarial losses or potential losses. Tax payers shouldn't have to subsidize people living in potential flood zones; insurance should be actuarially priced to reflect the risk and allow the private company to sustain themselves and earn enough of a profit to provide return for their investors as would any private business.

With subsidized contracts in the market, the way it is currently set up, means that the private company will be unable to compete with the national program that is willing to take on debt and deficit and pass it along to the tax payers, unlike the private industry that clearly cannot survive in that market. Therefore, the only way to get the private industry involved is to remove the subsidized contracts and have the contracts priced appropriately to reflect market conditions.

One of the biggest obstacles, Mr. Yun says, that stands in the way of this change is addressing the situation in which residents have resided on the property for a long time by paying the subsidized premium.

"There are many homeowners that purchased their property many years ago, and we do not want to shock these owners with dramatic increases in flood insurance premiums. It's just unfair for anyone living in a residence for such a long period of time to see a dramatic run-up in premium to the point where they may not be able to pay and will have to sell the property."

Currently, official NAR policy supports the privatization of the flood insurance program, but how that comes about is still up for debate. They are still deciding what is the necessary structure and platform to introduce the private market properly.

While actuarially sound insurance premiums are the long-term goal, we simply have to figure out what we can do now to get ourselves there in the future.

If the government suddenly decides to say that the flood program is continuing to run a deficit and we are leaving this market and no longer providing plans, then people will be left without subsidized insurance. Some people will take their own risks, others will buy the more expensive private insurance, and the attraction of the property value will no longer be there because of the increase in associated risk assessment and insurance costs.

This means there will be some type of price adjustment for those vulnerable properties. The high insurance premiums from the private market will cause the value of those properties to adjust downward.

In order to avoid a market shock like this, Mr. Yun recommends that the transition period should be somewhat gradual.

"Say a property should be paying $2,000, but is actually paying $500 in insurance costs," Mr. Yun described. "Then, how do we phase in the full cost? Over three years? Ten years? Or, should we account for how long somebody has lived in that residence?"

In coming up with a phasing solution for the private market, Mr. Yun and the NAR are tasked with the difficult process of combining social factors considering what is fair; they are also tasked with attempting to implement an actuarially sound and true insurance program rather than a subsidized one. Mr. Yun notes that a realtor's interest is to ensure their clients are fairly and reasonable treated, so they are really not trying to shock existing home owners with higher premiums.

Another approach to this issue involves focusing on a micro level, Mr. Yun explained.

"Let's say 10% of properties are within a flood zone, but maybe it's less than 1% that induces 85% of the flood costs. If it's just the most vulnerable 1% that causes the most damage and financial cost, then maybe we should just focus on these properties intensively and separate them out from the other properties for insurance purposes."

What Mr. Yun means by this is that the other properties would fall under the new private market with actuarial prices and the extreme risk prone properties would fall under a government program that explicitly says they will subsidize it for a number of years, maybe five or ten. Therefore, this will effectively integrate the private market to cover 90-95% of the properties in a flood zone without creating a price shock for those extremely vulnerable properties.

After listening to Mr. Yun's idea, I immediately liked what I was hearing. By subsidizing only for a certain period of time, that allows the owners time to save up money and invest in risk mitigating initiatives so that once their subsidized time is up, the premium shock will not be as drastic because they will have invested in risk mitigating initiatives that ultimately lower their actuarial premium. This phasing approach is a much better response than for the government to continue subsidizing forever the most at risk properties. With this approach, they are now incentivizing homeowners to fortify their homes before the subsidized contract runs out.

Regarding future development, a shift in policy like this means there will be less development in the highly risk prone areas. While this is often the reason cited for why not to privatize the process, Mr. Yun notes that there is always housing demand with populations continuing to increase, so there will be development in other areas and, in net, it doesn't hurt the real estate development industry.

Homes will be built further inland and, therefore, there will be more properties for sale inland, and only the truly vulnerable coastal market will suffer.

While it will hurt the vulnerable markets, that is its purpose: to dissuade people from continuing to develop unsustainably in markets that are at high risk for coastal storms.

The less risk prone areas will thrive and benefit. New Orleans is a prime example of when plenty of people said enough is enough and moved even without a privatized process. New Orleans has had a permanent outflow of people that have never really come back. Mr. Yun notes that the employment level is not what it was before Katrina, and similar effects may be expected for other highly at-risk communities.

CHAPTER 10

BID, RELI, AND A CALL TO ARMS

"It's tough to predict the future, but it's hard to imagine a future where there isn't more attention to long-term impacts of climate change and integrating them more effectively into CRE investment and development." Billy Grayson, Urban Land Institute Executive Director for the Center of Sustainability and Economic Performance, Bisnow 2018

So far, we analyzed the LEED certification program, we understand the private industry's rational behind not investing in risk mitigating activities, and we explored the inefficiencies associated with the NFIP in its current form. Through our market studies, we identified unique issues and solutions that can be applied to various other markets facing similar risks.

After looking at Houston, it is clear how a lack of zoning requirements can lead to over construction and result in significant damages in the aftermath of a massive storm. Moving forward, emphasis should be placed on updating storm water retention systems to minimize the economic impact that flooding has on vulnerable real estate markets.

In New York, it is clear Lower Manhattan should likely have never been developed to the extent it is today. However, now the challenge is about playing catch up and adequately protecting one of the most prized real estate markets and the world's financial hub. From Mr. Bowman, we learned about the difficult politics that slow the process of building a wall and the need for an integrated group effort. Also, Mr. Golden emphasized how desperate the state of our country's coastal infrastructure is and stressed the need to invest heavily in updating our ports in order to continue to grow and remain competitive in the age of globalizing economies.

We then looked at the three successful examples of storm water barriers in New England and examined the efforts already underway to protect the Boston Harbor for years to come. Climate Ready Boston is leading the way for a brighter future and should be referenced as a model approach for all other markets facing similar risks.

The research and insights gathered through conversations and interviews with experts in their respected fields has led me to summarize the key aspects that need our focus in order to make an impact mitigating future risks in coastal markets. In order to properly address the problem at hand, both short-term and long-term approaches are required.

SHORT TERM:

The role institutional investors play and the preferences they have for investments help guide the direction of the real estate industry. After hearing from Mr. Green in Chapter Five, we learned there are certain institutional investors with an interest in the problem at hand. Their investor preferences are a key strategic tool that must be used to help promote risk-mitigating initiatives.

Jomar Ereso, who is the director of asset management at RCLCO, a leading real estate advisory firm, said that "Institutional investors are increasingly asking the question and evaluating their potential exposures to climate-change-related impacts."

Pension funds like the State Teachers Retirement System of Ohio (STRS) have already jumped on board with this trend and are beginning to realize the costly trajectory of both the industry and the environment. STRS has a longer investment horizon, holding their properties for 20-30 years, making them much more exposed to the impacts of increasing storm surges and costs of damages associated with natural disasters.

"Every year, the average temperature increases across the world," says STRS Ohio Acquisition Director Eric Newberg. "You have to start paying attention to that when looking at real estate and making sure cities are addressing it or addressing assets correctly, or just decide not to invest in certain markets," says Newberg.

According to Newberg, their fund no longer invests in Florida after the insurance premiums jumped significantly following Hurricane Andrew in 1992, and haven't considered re-entering the market given the increasing risk of flooding.

Investors and real estate asset managers are and will continue to reassess their risks when investing in risky coastal communities. We are already beginning to see momentum and responsibility at the hands of long term investors, but difficulty lies with shorter term investors that may simply attempt to play a risky game of timing the next big storm. How climate change will impact commercial real estate moving forward remains a tough topic to talk about among investors, owners, and developers/operators because of the different time horizons associated with each practice.

Moving from investors to insurance, according to an interview with Bisnow in early 2018, Urban Land Institute Executive Director for the Center of Sustainability and Economic Performance Billy Grayson said,

"If you think that insurance is always going to cover you, you have to recognize insurance changes every year. When thinking about insurance rates and overall costs of running a building, it's going to change a lot in the next 50 years."

In the short term, adjusting insurance rates is one of the main ways we can begin to incentivize proper risk-mitigating development.

"Instead of insurance companies refusing to cover an asset or area, these companies should better incentivize property owners when they retrofit or build more resilient buildings," said Mr. Grayson.

A prime example of this type of behavior already taking form is at the mixed-use building *181 Fremont* in San Francisco, CA. In response to the increasing threat of earthquakes in the Bay area, "Structural engineer Arup added a shock absorption system to help it withstand a major earthquake. This system earned the building's owner, Jay Paul Co., favorable policy terms from its insurer and was one of the main reasons Facebook chose the building, according to Grayson."

This type of incentivizing behavior is great news for the industry and should be mimicked by other insurance companies operating in flood prone areas.

According to a 2018 Bisnow report, insurers like "AI reported single-digit rate increases during each of the months in Q4 2017, which has been sustained into Q1 2018. Regular property commercial offices in areas like Wisconsin where there were not disasters will see 0-5% rate changes, Bane Said. California is expected to have high-single-digit increases, while Florida might have 10-11% increases."

Through a combination of raising rates and incentivizing risk mitigating initiatives, the insurance industry can help reduce the future economic impacts natural disasters have in coastal real estate communities. Thinking more long-term insurance, the impetus is clearly on the government to reform the NFIP in its current form. Through a plan similar to that proposed by Mr. Yun in Chapter Nine, private insurance needs to be slowly phased into coastal markets in order for pricing to reflect accurately the amount of risk undertaken.

Additionally, the government can further incentivize small risk mitigating activities for developers like the ones Mr. Rutter mentioned in Chapter Seven through tax write-offs or similar subsidies for proactive behavior. With the billions of dollars flooding into disaster struck areas for relief, the government must properly allocate these funds and change their historical approach from one focused on restoration to a proactive approach focused on funding sustainability and mitigating future risks by incentivizing and promoting smarter development.

A tax write-off incentivizing safer and more sustainable development would be similar to a 401A tax policy for affordable housing. A 401A is basically a tax benefit that developers receive if they build a multifamily property in certain areas and have a specified percentage of that property be affordable housing for lower income brackets. By taking this approach and providing these benefits to the entire market, it will help individual developers realize some type of reduced cost or return on their investment for further mitigating their properties.

FEMA's flood maps should be updated to incorporate the impact of possible subsidence like that hypothesized in Houston and the increasing knowledge that surrounds the impact of climate change on coastal communities. It should project the likelihood of further implications in the future instead of using an historical facing analysis.

LONG TERM:

There are a number of initiatives that will certainly take longer to come to fruition, but it is imperative we begin addressing them immediately in order to effectively prepare for future storms.

Certain markets, including but not limited to those identified through the market studies such as New York and Boston, need storm surge barriers to protect the billions of dollars of real estate at risk.

How to finance these projects becomes the real question. A portion of the relief packages sent to devastated areas can help contribute to this as well as the catastrophic bonds proposed by Mr. Kirshen in Chapter Eight. Furthermore, another potential solution is a group effort, where those being protected and benefiting from a wall will help finance its construction.

This can be accomplished through a type of special BID tax that is focused on mitigating flood risk in vulnerable markets. Business Improvement Districts, also known as BID's, are organized and established by business and property owners. Each member of a BID pays a special BID tax that then goes towards organized efforts to enhance the economic vitality of downtown commercial areas by focusing on things like improving landscaping, reducing trash and graffiti, increasing security presence and making capital improvements such as street benches and lighting. There are ten different BID's in D.C. alone, and I know these organizations are already formed in other cities like New York and Boston. It would not be terribly difficult to use these pre-existing organizations to help organize the necessary funding to pursue risk mitigating activities.

MODELING LEED:

Returning back to our discussion of LEED, we learned that the program promotes good practices in the construction and development process by reducing the carbon footprint of projects and the impact construction has on the environment. However, the long-term benefits of adopting LEED are not yet quantifiable because it is still in an early stage of adoption. From Ms. Nascimento, we learned that the adoption of LEED has been slow for some because they have had trouble realizing a return on investment from LEED efforts. Until we are able to demonstrate that the cost benefit of mitigating flood risk is less expensive than simply rebuilding or passing costs on to the private insurers, then adoption of flood-risk mitigating initiatives will also be slow.

In Chapter Five, however, we learned that currently the greatest benefit realized from adopting LEED is its use in marketing properties. Specifically, in residential and commercial office real estate, the LEED initiatives resonate well with customers and consequently, foreign investors as well.

Both architects and engineers are training in LEED so there is clearly a focus from the industry on the future of the program. While the initiatives of LEED don't necessarily promote flood-risk mitigating actions, through observing the techniques LEED has used to gain attention and adoption in the real estate industry, we can learn and model how to approach flood-risk mitigating initiatives in a similar manner.

That is exactly what is being attempted with the implementation of RELi.

Back in 2012, work began on RELi, a certification process that will revolutionize the way we address the relationship between development in coastal cities and climate change. The work has been led by the Market Transformation to Sustainability (MTS), an American National Standards Institute (ANSI) accredited standard developer, in collaboration with global architecture firm Perkins+Will, Eaton Corporation, Deloitte Consulting, and Impact Infrastructure.

The program will aim to promote sustainable development and guide all aspects of the real estate process, from architects to developers and city planners to local governments, to better prepare for all types of natural disasters, including but not limited to wild fires, hurricanes, earthquakes, and severe coastal flooding. It will help push construction processes in coastal communities to be proactive in their response to climate change and mitigate risks through adaptive design for rising sea levels, increased storm surges and excessive rain.

While the majority of money directed at natural disasters is focused on simply restoring communities to pre-storm conditions, even more will be required to mitigate future risks. "Resilience planning and adaptation are some of the most expensive public and private activities in U.S. history," says Mike Italiano, president and CEO of MTS. Mr. Italiano further highlights the scale of the problem we are trying to fight when mentioned in a publication for Perkins+Will:

"Independent research shows that the U.S. has tens of trillions of dollars in resilience costs. S&P and Moody's are issuing climate credit rating downgrades for entities that are not resilient, and the insurance industry is withdrawing from high-risk markets because the industry considers climate change 'an uninsurable risk.' When applied to buildings, homes, and infrastructure, our national consensus standard, RELi, can greatly reduce these unprecedented costs and risks."

The shift in these credit rating standards is a great step towards focusing on sustainability and resiliency of vulnerable markets in the future. The RELi program will hopefully work to help these markets think smarter about development and mitigate risk to align with the new standards set forth by the credit agencies.

On November 8th, 2017, the U.S. Green Building Council adopted RELi, providing it with the resources and governing body already utilized by LEED. In reflecting on this large milestone achieved after years of hard work, Doug Pierce, a principal investor for RELi and co-director for Perkin+Will's Resilience Research Lab, said, "we are delighted, if not humbled, that the USGBC is embracing this innovative tool that we and our partners have worked on so diligently and so passionately for the last half-decade."

Thanks to Perkins+Will, and all of the other firms that have taken the initiative and have been involved in the creation of this program, the industry now appears to be heading in the right direction. "What we're seeing now is the merging of thought leadership from some for the world's most progressive designers and thinkers with the global organizing capacity of the USGBC. It's going to create unprecedented potential for market transformation toward resilience planning and resilient design."

Through my research and interviews with large private real estate developers, it is clear the industry has been lacking the appropriate incentives and leadership to focus on resilience to climate change. Now, the focus is on taking this initiative and turning this potential for market transformation into tangible results. To help this happen, RELi has once again looked to its big brother, LEED, for guidance. RELi has created an extensive training program to certify individuals as RELi Accredited Professionals (APs), similar to LEEDs APs. Administered by MTS, the program ensures that every AP is a master of resilient design. As of November 2017, there are 12 AP's worldwide, all of which are employees of Perkins+Will.

Currently, oversight from reliable AP's and implementation of RELi's standards are being used in a number of different projects around the country. Now, following the adoption by the USGBC, the number of properties implementing this program is expected to grow significantly. One of these current projects includes the Christus Spohn Hospital located in Corpus Christie, TX, where Harvey first made landfall as a Category 4 storm. Surprising to some but not to those internal to the program, the hospital was able to withstand Hurricane Harvey with little to no damage. This further supports that the initiatives set forth by this program are making a material impact on coastal communities, reducing the costs of damage and loss of life in these vulnerable areas.

It is a pilot project like this that will help pave the way for future sustainability and risk mitigation initiatives. The effectiveness of and assessed impact on these pilot projects will help structure the rating and point system for RELi in order to be a fully functioning certification process. While this is only the beginning, in order to lead RELi into the future, a Resilience Steering committee, similar to LEED's Steering

Committee, has been created and will be led by Doug Pierce of Perkins+Will.

The goal of this committee will be to continuously asses the risks in coastal communities and refine the standards set forth in RELi's certification process moving forward. Mirroring the LEED program in many ways, the comprehensive approach of RELi is a response to the fact that "our communities, our neighborhoods, and our buildings are all interconnected," said Janice Barnes, Global Resilience Director and Co-Director of the Resilience Research Lab at Perkins+Will. The future of these coastal communities will now rely on "hazard preparation and adaption as well as chronic risk mitigation" set forth through the RELi program.

Lastly, I believe that one day, risk mitigating projects and initiatives could be used as a tool when sourcing capital and trying to attract investors that value sustainable development.

Mr. Bhatia of ASB Capital, who was introduced back in Chapter Three, concluded by saying, "Ultimately, however it's done, mitigating risks does provide a safety net for the property, and, if people want to feel safe in markets like Miami, Boston, or New York, then it could be a differentiator to have."

Currently, however, Bhatia and the rest of the industry are not seeing that type of demand from the market or institutional investors. There is not a premium in most tenants' minds for mitigating flood risks like there is for LEED certifications that focus on energy efficient operation.

Well, why is that? I think that's a question we all should be asking ourselves.

In order to create a lasting impact and save the coastal real estate markets we all enjoy, we must begin to acknowledge their inherent risks and value their safety moving forward. After reading this book, you are armed with the knowledge and insight to promote risk-mitigating initiatives and influence the value we place on certain at-risk properties. It is up to you, using not only your voice but your dollar as well, to fight for sustainable and smarter developed coastal communities.

Imagine the regional stability and economic prosperity that will come from an investment in our coastal infrastructure. Most importantly, imagine the peace of mind that will come knowing your community is safe and protected in the face of the next storm.

ACKNOWLEDGEMENTS

Throughout the process of writing this book, many individuals have taken time out of their day to help contribute to this piece. Thank you to all my close friends and relatives who continuously asked me how the process was going and unknowingly motivated me to keep working and ultimately complete this book. Thank you to the faculty from the Steers Center for Global Real Estate and the other industry alumni, who provided me with connections and contacts needed to develop the insights provided in this book. I would also like to specially thank my Uncle, Curtis Wahle, for fostering my interest in real estate from such a young age. Your leadership, support, and guidance in life and in writing this book is greatly appreciated. Thank you to all of my past colleagues at TTR Sotheby's, Vine Street Studios, Morgan Stanley, and Gilbane Development for providing me the opportunity to continue exploring my interest in the real estate industry. Thank you to my editors and publisher working day and night to finalize this project on schedule. I would also like to give a special thanks to my high school advisor, Tyler Caldwell, for providing his professional advice and assistance in polishing this manuscript. You have been an incredible mentor and I truly cherish our friendship. Lastly, I would like to thank my parents, Rory and Betsy, and my siblings, Amanda and Everett, for providing your support and encouragement throughout my life as I continue to pursue my own personal goals.

REFERENCES

Book Description

"National Report." *The Economic Risks of Climate Change in the United States—Risky Business*, 2014, riskybusiness.org/report/national/.

Doggett, Tim. "The Growing Value of U.S. Coastal Property at Risk." *AIR Worldwide*, 23 Apr. 2015, www.air-worldwide.com/Publications/AIR-Currents/2015/The-Growing-Value-of-U-S--Coastal-Property-at-Risk/.

HymanOct, Randall. "New York City Area Could Soon See Massive Floods Every 5 Years." *Science Magazine*, 8 Dec. 2017, www.sciencemag.org/news/2017/10/new-york-city-area-could-soon-see-massive-floods-every-5-years.

Introduction

Allen, Karma, and Maia Davis. "Hurricanes Harvey and Irma May Have Caused up to $200 Billion in Damage, Comparable to Katrina." *ABC News*, ABC News Network, 11 Sept. 2017, abcnews.go.com/US/hurricanes-harvey-irma-cost-us-economy-290-billion/story?id=49761970

Doggett, Tim. "The Growing Value of U.S. Coastal Property at Risk." *AIR Worldwide*, 23 Apr. 2015, www.air-worldwide.com/Publications/AIR-Currents/2015/The-Growing-Value-of-U-S--Coastal-Property-at-Risk/.

Nichols, Wallace J. "BLUEMIND: This Is Your Brain on Ocean." *The Huffington Post*, TheHuffingtonPost.com, 25 May 2011, www.huffingtonpost.com/wallace-j-nichols/bluemind-brain-ocean_b_863986.html.

Sneed, Annie. "Hurricane Irma: Florida's Overdevelopment Has Created a Ticking Time Bomb." *Scientific American*, 12 Sept. 2017, www.scientificamerican.com/article/hurricane-irma-floridas-overdevelopment-has-created-a-ticking-time-bomb/.

Worland, Justin. "Hurricane Harvey: Bad Policy Left Houston Unprepared." *Time*, Time, 29 Aug. 2017, time.com/4919224/hurricane-harvey-houston-policy/.

Worland, Justin. "South Carolina Flooding: Climate Change Causes Worse Storms." *Time*, Time, 5 Oct. 2015, time.com/4061371/south-carolina-flooding-climate-change/.

Chapter 1: The Extent of the Problem

Astor, Maggie. "The 2017 Hurricane Season Really Is More Intense Than Normal." *The New York Times*, The New York Times, 19 Sept. 2017, www.nytimes.com/2017/09/19/us/hurricanes-irma-harvey-maria.html.

Burleigh, Nina. "New York City Seen Becoming Atlantis If It Doesn't Spend Billions of Dollars on Seawalls." *Newsweek*, 30 Oct. 2017, www.newsweek.com/hurricane-sandy-new-york-rising-sea-level-floods-695288.

Drye, Willie. "2017 Hurricane Season Was the Most Expensive in U.S. History." *National Geographic*, National Geographic Society, 30 Nov. 2017, news.nationalgeographic.com/2017/11/2017-hurricane-season-most-expensive-us-history-spd/.

Gray, Sarah. "Boston Is Experiencing Flooding from Nor'easter Storm Riley." *Time*, Time, 2 Mar. 2018, time.com/5183334/boston-flooding/.

HymanOct, Randall. "New York City Area Could Soon See Massive Floods Every 5 Years." *Science Magazine*, 8 Dec. 2017, www.

sciencemag.org/news/2017/10/new-york-city-area-could-soon-see-massive-floods-every-5-years.

Kaplan, Thomas. "Senate Approves $36.5 Billion Aid Package as Hurricane Costs Mount." *The New York Times*, The New York Times, 24 Oct. 2017, www.nytimes.com/2017/10/24/us/politics/senate-congress-disaster-aid-hurricanes-fire.html.

Littman, Julie. "Real Estate Investors Continue To Roll Dice On Disaster-Prone Assets." *Bisnow.com*, 18 Feb. 2018, www.bisnow.com/national/news/capital-markets/real-estate-investors-continue-to-roll-dice-on-disaster-prone-assets-85118.

"National Report." *The Economic Risks of Climate Change in the United States—Risky Business*, 2014, riskybusiness.org/report/national/.

Struyk, Ryan. "What Past Federal Hurricane Aid Tells Us about Harvey." *CNN*, Cable News Network, 7 Sept. 2017, www.cnn.com/2017/08/31/politics/hurricane-harvey-recovery-money/index.html.

Chapter 4: Good Fast Cheap

Marso, Jaime. *Natural Hazard Mitigation Association*, NHMA, nhma.info/

Chapter 5: Institutional Interest

LEED | USGBC, new.usgbc.org/leed.

Chapter 6: Houston We Have a Problem

Cappucci, Matthew. "Hurricane Harvey's Flood Threat Sparks Memories of Tropical Storm Allison in Southeast Texas." *The Washington Post*, WP Company, 24 Aug. 2017, www.washingtonpost.com/news/capital-weather-gang/wp/2017/08/24/hurricane-harveys-flood-threat-sparks-memories-of-tropical-storm-allison-in-southeast-texas/?utm_term=.3b6abda761c1.

Harden, John D. "Breaking down Houston's Recent Flooding Events." *Houston Chronicle*, Houston Chronicle, 27 Apr. 2016, www.houstonchronicle.com/local/article/How-floods-compare-7330750.php.

KHOU.com. "Buffalo Bayou to Remain at Record Level; Barker, Addicks Reservoirs Have Peaked." *KHOU*, 1 Sept. 2017, www.khou.com/weather/hurricanes/hurricane-harvey/controlled-release-of-barker-addicks-reservoirs-to-impact-thousands/468348109.

"Privacy Policy." *AEP Texas Hurricane Harvey Restoration Update—9-3-2017, 4:30 P.m.*, 3 Sept. 2017, www.aeptexas.com/info/news/viewRelease.aspx?releaseID=2339.

Chapter 7: New York is Worth Saving

"Building the Case for Coastal Resilience." *National Institute for Coastal Harbor Infrastructure*, www.nichiusa.org/.

Farberov, Snejana. "How Hurricane Sandy Flooded New York Back to Its 17th Century Shape as It Inundated 400 Years of Reclaimed Land." *Daily Mail Online*, Associated Newspapers, 16 June 2013, www.dailymail.co.uk/news/article-2342297/Manhattans-original-coastline-revealed-Hurricane-Sandy-flooded-land-reclaimed-400-years.html.

Fischetti, Mark. "Russian Flood Barrier Is a Model for New York City." *Scientific American*, 10 June 2013, www.scientificamerican.com/article/russian-flood-barrier.

Gifford, Victor. "George Washington Wept Here (Part III): How A Tiny New Jersey Town and a Korean Conglomerate Are Set To Despoil An American Treasure." *Wild's Brigade*, wildsbrigade.blogspot.com/2014/05/george-washington-wept-here-part-iii_26.html.

Littman, Julie. "Real Estate Investors Continue To Roll Dice On Disaster-Prone Assets." *Bisnow.com*, 18 Feb. 2018, www.bisnow.com/national/news/capital-markets/real-estate-investors-continue-to-roll-dice-on-disaster-prone-assets-85118.

"Posts about Calgary Flood on Xraydelta." *Xraydelta*, xray-delta. com/tag/calgary-flood/.

Sanderson, Eric. "New York—before the City." *TED: Ideas Worth Spreading*, July 2009, www.ted.com/talks/eric_sanderson_pictures_new_york_before_the_city.

"The Welikia Project » Mannahatta Curriculum." *The Welikia ("Way-LEE-Kee-Uh") Project*, welikia.org/download/curriculum/.

Chapter 8: Boston Pushes Forward

Celano, Lee. "As Seas Rise, City Mulls a Massive Sea Barrier across Boston Harbor." *BostonGlobe.com*, 18 Feb. 2017, www.bostonglobe.com/metro/2017/02/18/seas-rise-city-mulls-massive-sea-barrier-across-boston-harbor/dxtlbGrfSmYE2zacwUKakJ/story.html.

Kourakis, Yianni, and Mark Dondero. "Ceaseless Repairs Keep Fox Point Hurricane Barrier Holding Strong." *WPRI 12 Eyewitness News*, 11 July 2016, wpri.com/2016/07/11/ceaseless-repairs-keep-fox-point-hurricane-barrier-holding-strong/.

"NEW ENGLAND DISTRICT." *New England District*, www.nae.usace.army.mil/Missions/Civil-Works/Flood-Risk-Management/Connecticut/Stamford-Hurricane-Barrier/.

"NEW ENGLAND DISTRICT." *New England District*, www.nae.usace.

army.mil/Missions/Civil-Works/Flood-Risk-Management/ Massachusetts/New-Bedford/.

"NEW ENGLAND DISTRICT." *New England District*, www.nae.usace. army.mil/Missions/Civil-Works/Flood-Risk-Management/ Rhode-Island/Fox-Point/.

Chapter 10: BID RELi and a Call to Arms

Littman, Julie. "Real Estate Investors Continue To Roll Dice On Disaster-Prone Assets." *Bisnow.com*, 18 Feb. 2018, www.bisnow.com/national/news/capital-markets/real-estate-investors-continue-to-roll-dice-on-disaster-prone-assets-85118.

Quinlan, Ryan. "U.S. Green Building Council Adopts Resilient Building and Design Standard 'RELi'." *PerkinsWills.com*, 22 Nov. 2017, perkinswill.com/news/us-green-building-council-adopts-resilient-building-and-design-standard-reli.

www.ingramcontent.com/pod-product-compliance
Lightning Source LLC
Chambersburg PA
CBHW030925180526
45163CB00002B/471